APPLICATIONS OF VHDL TO CIRCUIT DESIGN

APPLICATIONS OF VHDL
TO
CIRCUIT DESIGN

edited by

Randolph E. Harr
C.A.D.onomist

Alec G. Stanculescu
Fintronic USA, Inc.

Kluwer Academic Publishers
Boston/Dordrecht/London

Distributors for North America:
Kluwer Academic Publishers
101 Philip Drive
Assinippi Park
Norwell, Massachusetts 02061 USA

Distributors for all other countries:
Kluwer Academic Publishers Group
Distribution Centre
Post Office Box 322
3300 AH Dordrecht, THE NETHERLANDS

Consulting Editor: Jonathan Allen

Library of Congress Cataloging-in-Publication Data

Applications of VHDL to circuit design / edited by Randolph E. Harr,
  Alec G. Stanculescu.
    p. cm.
  Includes bibliographical references and index.
  ISBN 0-7923-9153-5
    1. Electronic circuit design—Data processing. 2. VHDL (Computer
hardware description language) I. Harr, Randolph E., 1959–
II. Stanculescu, Alec G.
TK7867.A64 1991
621.381 '5—dc20   91–10756
                                                        CIP

Contents

Table of Figures

Foreword

Describing and designing complex electronic systems has become an overwhelming activity for which VHDL is showing increasingly useful and promising support. Although created as a description language, VHDL is being increasingly used as a simulatable and synthesizable design language. For the first time, here is a book which describes a number of unique and powerful ways VHDL can be used to solve typical design problems in systems -- ones which must be designed correctly in very short periods of time. Typically useful techniques such as switch-level modeling, mixed analog and digital modelling, and advanced synthesis for which VHDL shows great promise are fully presented. These methods are both immediately applicable, and indicate the potential of VHDL in efficiently modelling the real world of electronic systems.

Since its inception, there has been a desire for an analog description language consistent with (and integrated with) VHDL. Until recently, VHDL could only be applied to digital circuits, but the dream of describing and simulating mixed analog and digital circuits is now a reality as described herein. Describing the functionality of analog circuits including interoperability with digital circuits using the VHDL paradigm is surprisingly easy and powerful. The approach outlined by the authors presages a significant advance in the simulation of mixed systems.

The efficient automatic design of digital circuits is becoming increasingly important as well. Such capabilities are rising from the research and being applied to the industrial environment. Challenges still exist, but much is known about how to model systems such that efficient synthesis can be done. The approach to synthesis modelling as outlined here is useful for anyone contemplating using VHDL for synthesis.

A basic premise of VHDL is to describe digital components consistent throughout all levels of descriptive abstraction - gates to systems. Just as hardware components such as standard cells, chips, boards, or entire systems are considered as standard component parts for the construction of new assemblies and systems, there must be a synonymous and consistent way to describe these parts as reusable components in the VHDL environment. A methodology for constructing VHDL component libraries is presented here.

Finally, VHDL is a recent standard and its problems are becoming increasingly understood. A treatise on what the more significant problems are, and how to work around them, is important for every designer to know. Such a treatise is important for all designers and included in this book.

The authors of this book have attempted, with great success, to present modern methods, capabilities, and extensions to the already well-known abilities of VHDL to serve the electronic systems industry. This is a book to be devoured, followed, and used by every electronic system designer.

Dr. Harold W. Carter
University of Cincinnati
Cincinnati, Ohio

Preface

VHDL is an industry standard modeling language for the design and description of electronic systems. Although intended for digital hardware design, VHDL contains features found in modern programming, hardware description, and C.A.E. design and simulation languages. This rich feature set is allowing the language to be adopted by a larger and more diversified audience than for any other hardware design / description language (HDL). It is for this audience of users that this book is written.

When the initial idea for this book developed in early 1989, it was envisioned to cover the complete design space from transistor circuit to queueing theory system models. It quickly became evident that to do any topic justice required more space and time than was available from a single author. Hence, the book expanded to contributed chapters in a series of books entitled "Applications of VHDL to ...". This initial book, "Applications of VHDL to Circuit Design", focuses on circuit modeling and timely topics about VHDL's use in the electronic design process.

The book series intends to bridge the gap between electronic design and the VHDL language standard. Concepts difficult to model in the electronic design process are introduced with solutions in VHDL. Designers familiar with VHDL can start with the examples and explore further solutions to their unique problems. Those not familiar with VHDL can adapt the complete examples without a detailed understanding of the language. The book series does not attempt to serve as an introduction to the VHDL language but rather concentrates on providing solutions to real-world modeling problems.

There is real irony in that circuit design is the first topic covered in the book series. When designed in the early 1980's, VHDL was targeted for logic and digital function description only. Techniques in VHDL of discrete-event driven simulation and hierarchical modeling were not perceived as applicable to analog and physical circuit models. And yet due to these features and others, VHDL is being easily adapted to circuit and mixed level modeling.

Along with the circuit design focus, the book introduces the reader to VHDL application areas such as synthesis and standard parts libraries. In the editors experience, understanding the synthesis of gates from VHDL is the best way for a today's hardware designer to learn the features of VHDL. These application areas will be expanded on in future volumes.

The closing chapter in the book highlights common misunderstandings in the current language. The 1992 re-standardization effort underway is the opportunity to correct these ideas and coalesce new ones conceived during language use. You are encouraged to inject any developed knowledge into the standardization process so VHDL will remain a technically useful standard. Contact the IEEE or the VHDL Users' Group to learn how.

ELECTRONIC DISK SUPPORT (Reprint edition)

To further support and promote the application of VHDL, the authors have donated their models from the book into the public domain. Note that donating models into the public domain has not removed the copyright thereof. See the notices on the electronic material and at the front of this book concerning any material printed in this book or contained on the electronic files originally distributed freely with the book but now available on the Internet. An electronic copy of all the donated VHDL code shown or referenced in the chapters has been compiled. Kluwer Academic Publishers, in wishing to further support the VHDL community, graciously agreed to distribute this electronic copy on a floppy diskette free of charge with each copy of the original printing of the book. Luckily, with today's internet age, the material can now be more easily and reliably distributed via an HTTP/FTP server .Retrieve the files from http://www.eda.org/misc/appcircVHDL.zip or ftp://ftp.eda.org/misc/appcircVHDL.zip See the README file in the zip archive for information on the contents and file formats. Kluwer Academic Publishers is not supporting the electronic copy of material.

Randy Harr, Los Altos Hills, CA (Jan 2001)

---

[1]   Note that donating models into the public domain has not removed the copyright thereof. See the notices on the electronic disk and at the front of this book concerning any material printed in this book or contained on the electronic disk distributed freely with the book.

BECOME A CONTRIBUTOR

As mentioned earlier, there will be a series of books entitled "Applications of VHDL to ...". Even if you never considered writing before, this becomes a good opportunity to develop your ideas in detail for the benefit of others. A single chapter is much easier and quicker to develop than a full manuscript. Encouraged are submissions from engineers out in the field applying VHDL to real design problems. If you have interesting ideas on using VHDL in the design process, then submit a draft outline or manuscript as a proposed chapter. Send it to the Editor care-of the before listed address.

As a selected author, you will retain copyright to any material you contribute and share in the success of its publication. Authors should expect to eventually develop a final, camera ready copy of their material in the same style and format as this work. Don't delay; submit today!

ACKNOWLEDGEMENTS

Any undertaking of a large amount of written material takes the dedication and help of many people. Without their invaluable assistance such projects never come to fruition. Therefore, the editor and authors wish to recognize the people who played an important role in bringing this work to print.

Randy Harr wishes to thank the authors for their patience and understanding during the extended period over which this work developed. Many thanks to Carl Harris of Kluwer who has been extremely patient and supportive during this process. Personal support from his wife Roshan during the many hours of work has been greatly appreciated. Checking final copy on their honeymoon was definitely an early test of "in good times and in bad". Credit is due Alec Stanculescu for helping promote the idea of the book and assisting with the development. For his additional work he is merited with being listed as Co-editor on the cover. Many thanks to "Bob" Stanisic who arrived with more chapters just as other authors were late with their material! Look for more work from this promising young author in the future.

Alec Stanculescu wishes to thank Andy Tsay, Alex Zamfirescu, and Doug Perry who contributed to [9] and [10] -- the basis for his chapter. Many thanks to Vantage Analysis Systems for placing the 46-value system and the accompanying package in the public domain. Finally, thanks to all the Vantage employees, and to David Coelho and Rick Lazansky in particular, for their constant support.

Bob Stanisic wishes to thank his wife Margie for the support and patience during the long hours of work. Bob and Mark wish to thank Larry F. Saunders and Robert L. Hedman for reviewing their work.

Steve Carlson writes that "the development of [his] material was a true team effort. Many of the ideas and examples presented come from the core of the Synopsys VHDL [synthesis] product development team. Many thanks to those team members: Bill Krieger, Russ Segal, Brent Gregory, Jerry Huth, and Emil Girczyc. The diligence of the many others from the Synopsys team who participated in early reviews is greatly appreciated. Special thanks are extended to Janet Greene who proved to be tireless in handling the revisions and layout of the work."

Gabe Moretti wishes to thank Vantage Analysis Systems, Inc. for providing their VHDL Spreadsheet environment; and to extend thanks to Gary L. Mullin for typesetting the manuscript.

Last, but by no means least, Ken Scott wishes to acknowledge Synopsys for support of his work. Also, he extends thanks to Brent Gregory who provided detailed feedback on a near-final draft. Finally, he wishes to acknowledge his wife, Shelly, who has been very understanding and supportive throughout the effort.

Finally, thanks to you the thousands of electronic system developers out there who constantly challenge old methods in hopes of expanding the frontiers of the electronic design process.

Randolph E. Harr
San Francisco, California

APPLICATIONS OF VHDL TO CIRCUIT DESIGN

Chapter 1

Switch-Level Modeling in VHDL

Dr. Alec G. Stanculescu [1]

Fintronic USA, Inc.
Menlo Park, California

1.1   Introduction

VHDL is a language to build and program an event-driven simulator. For
this purpose it is excellent. It is even better than it was intended to be!
For example, even the designers of the language considered the behavior
of the bi-directional pass-transistor outside the scope of pure VHDL. It
is in part to their credit that they were wrong on this issue. VHDL is a
very powerful language for programming an event-driven simulator.

This chapter intends to present advanced programming capabilities
available in VHDL and the topic of switch-level modeling. The reader
will become familiar with advanced VHDL features, while being kept

---

[1]Sections of this chapter have been reprinted, with the permission of the IEEE,
from the paper "Switch-Level VHDL Descriptions", published in the proceedings of
ICCAD89 [9]

motivated by solving a useful and challenging problem. This material is based on work initially presented at ICCAD89 [9], and on a 46-value system presented at a VDEG meeting [10].

This chapter, as well as the others in the book, assumes that the reader has a basic knowledge of VHDL. This could have been obtained by reading an introduction to the language, such as [13], or going through an introductory course. The aim of this book is to acquaint the reader with using VHDL in modeling circuits.

1.1.1 Overview

After reviewing the basic problems and the motivation for switch-level simulation, Section 1.2 presents advanced VHDL features necessary for switch-level descriptions.

Next, section 1.3 introduces the 46-value system and the public domain standard_logic" package.

Section 1.4 presents a distributed algorithm of a bi-directional pass-transistor. The VHDL implementation of this algorithm is covered in section 1.5.

Section 1.6 presents examples of networks of bi-directional pass-transistors. The examples are intended to clarify the concepts introduced in the earlier sections. They also provide information regarding the actual simulation performance. Finally, Section 1.7 presents the status of the research in this domain, and points to future directions in switch-level and analog simulations.

1.1.2 Why Switch-level VHDL Descriptions

The design of MOS technology electronic circuits requires functional-level simulations, as well as switch and circuit-level simulations. Functional simulations are necessary to understand the behavior independently of the implementation details.

Switch-level simulation is used when detailed timing information is important to verify a correct implementation. Switch-level models are developed when the corresponding SPICE-like simulation is not practical. A survey of switch-level simulators and algorithms can be found in [3].

The integration of switch-level with functional simulation is difficult. The state-of-the-art still accepts different simulation runs for each. VHDL represents a suitable modeling language for both switch-level and functional simulations. This is true mixed-mode simulation, in the sense that functional components and switch-level components can be described in the same modeling language and communicate with each other during the same simulation run.

Some VHDL simulators already associate switch-level semantics to simulator primitives in order to add switch-level capabilities. [1]. The disadvantage of these implementations is that they are not "pure VHDL". Descriptions that run on one simulator with hard-wired switch-level primitives may not run on another full-VHDL simulator.

This chapter shows switch-level semantics described in pure VHDL. It provides the VHDL architecture of a bi-directional pass-transistor, and discusses the modeling capabilities of the std_logic package that can be used to model switch-level primitives.

Switch-level models described completely in VHDL legitimize switch-level support provided by VHDL simulation kernels. This ensures that full VHDL simulators will be able to run switch-level descriptions that are completely written in VHDL. This results in portability across platforms, as well as in having VHDL span behavior, dataflow, structure, gate, and switch-level descriptions.

1.1.3   A Switch-level Modeling Solution in VHDL

As in all simulation problems, the overall goal of switch-level simulations is to produce accurate and useful predictions of the behavior of the circuit in a short time. In developing a VHDL switch-level modeling methodology several problems were encountered:

1. Algorithms were not available in the public domain in sufficient detail to use them easily.

2. In order to obtain useful results, the simulation algorithm should not be overly conservative. It is useless to predict that all values will be unknown! Handling the unknown values in a manner in which information is not lost during simulation is a difficult task of great importance.

3. It was questioned whether VHDL had the programming features necessary to implement a distributed algorithm for bi-directional switches.

These problems were overcome with the developed solution.

1.1.4   Choice of Algorithm: Interpretative vs. Compiled, and Global vs. Distributed

Switch-level algorithms can have interpreted and/or compiled characteristics. The interpreted algorithms can be global or distributed (local). Choosing the right algorithm was crucial to its implementation in an event-driven simulation environment.

The difference between interpreted and compiled is not as clear-cut as many people may believe. We say that algorithm A is compiled and B is interpreted if a greater portion of B can be used unchanged for all circuits. We typically call the portion common to all circuits *the interpreter*. It reads the data for a given circuit and interprets it. As the time spent in the interpretative part becomes smaller, the algorithm is said to have a bigger compiled component. Therefore, it should come as no surprise to see algorithm B called interpretative when compared with A, and called compiled when compared with C.

A global algorithm has at each step access to the data for the entire circuit. A distributed algorithm has access only to the data associated to one component of the circuit.

Generally, interpretative algorithms are slower than the compiled ones, and the distributed algorithms are slower than the global ones.

The interpretative algorithms are easier to program than the compiled algorithms, typically because the compiled algorithms require an additional level of indirection in generating the code: one has to write code that should generate the code that will implement the algorithm. In the case of the interpretative algorithm, one has to write the code that implements the algorithm.

Any VHDL simulation environment supplies an event-driven simulator. Such a simulator can be used to interpret VHDL descriptions. Therefore it is easier to implement an interpretative algorithm for VHDL switch-level simulations.

A distributed algorithm has a lot of appeal because of a simpler integration in the VHDL environment. In VHDL many entities get instantiated and communicate with each other via ports of a given data type. Global communication is more restricted. Therefore, a distributed solution becomes the best choice.

It follows that the best choice for the algorithm is to be interpretative and distributed. Let us review the remaining difficulties encountered in modeling bi-directional pass-transistors in VHDL.

A bi-directional switch is either on or off. When it is on it should behave as a wire, and when it is off it should behave as a disconnection.

In a distributed algorithm, the model of an on switch must pass information from drain to source and vice-a-versa until the two values are the same. Handling loops of switches becomes a concern regarding the possibility of computational oscillations (e.g., the nodes in the real circuit stabilize to some values, whereas the values in the model keep oscillating at a given time point).

The problems associated with modeling the bi-directional switch are:

1. Can we define a value-system and a distributed algorithm so that the instances of bi-directional switch models communicating with each other do produce the expected simulation results ? Such implementations existed before, but not in the public domain.

2. Can we code such a distributed algorithm for the bi-directional switch in VHDL? This was considered impossible by experts.

Sections 1.3, 1.4, and 1.5 will give positive answers to these questions.

Meanwhile, the next section will review the VHDL features that make it possible to successfully program the VHDL event-driven simulator. Readers who are not familiar with VHDL should read this section with help from an introductory document on VHDL.

1.2   Advanced Simulator Programming

This Section presents the features of VHDL that support the programming of the event-driven simulator. In particular we focus on features that have been used in the implementation of the distributed algorithm for the bi-directional switch.

To understand a programming language in general, one has to understand three main topics:

1. An engine to be programmed, (e.g., the CPU in the case of assembly languages, the sequential state machine for procedural languages such as C, or the Simulation Cycle in the case of VHDL).

2. Special data containers associated to the engine, (e.g., special registers in the case of assembly languages, or predefined attributes in the case of VHDL).

3. A facility to define and use the data containers.

In the next sections we will discuss these three topics in relation to VHDL.

1.2.1   VHDL Simulation Cycle

The VHDL Simulation Cycle is performed by the simulation kernel (the interpreter part of a VHDL simulator). The kernel interprets the data associated to the circuit, and up-dates that data based on the description of the various instances of modules (which are transformed into corresponding processes).

At the beginning of the simulation, before the first simulation cycle, there is an initialization step which consists of initializing each signal or generic parameter with a value according to a set of rules prescribed by the IEEE 1076 VHDL Language Reference Manual [12]. Then all processes are executed once. With these two steps the initialization is completed and the first simulation cycle is ready to be executed.

In general terms, the VHDL simulation cycle consists of the following steps:

1. Go to next time point and process all transactions scheduled for that time point. If there are no transactions scheduled then the simulation is done. If there are scheduled then place the values in the appropriate drivers (contributions).

2. Evaluate signals (including guards). For each signal for which there was any actvity by one of its contributors (drivers), reevaluate the signal (invoking the resolution function if necessary).

3. Execute processes that are sensitive to signals that changed their values or wake up due to a timeout condition becoming true.

4. Go to step 1.

For a detailed description of the full cycle the reader should consult [12].

Note that a signal assignment, by default is scheduled for "0 ns" later. This represents one simulation cycle later, but not a later simulation time point! There can be an unlimited number of simulation cycles within any given time point.

A delay in VHDL is associated with every signal assignment. That is assignments to signals are scheduled differently based on the particular assignment. Other languages are more restrictive, in that they only allow the association of delays to the wires and not to drivers of the wire.

Note that any assignment with a non-zero delay may affect a signal only in the first delta (simulation cycle) of a time point. This property is used in the VHDL implementation of the bi-directional switch presented in Section 1.5.

1.2.2   Variables vs. Signals & Predefined Attributes

In VHDL, data is stored in signal, variable, or constant objects. Variables are mere containers of data, whereas signals have more semantics associated with them. Constants are compile-time place-holders for pre-simulation calculated values.

The differences between variables and signals can be summarized as follows:

1. The result of a variable assignment is the immediate change of the value stored in the variable, whereas the result of a signal assignment is placing in the prediction queue of a prediction of a contribution to the signal. It is a prediction because it may be preempted by a subsequent assignment, and it is the prediction of a contribution because there could be other contributions to the signal and the actual value of the signal is established by the resolution function.

   Note also that a driver may stop driving a signal if it corresponds to a guarded assignment and the guard becomes false.

2. A signal has associated with it more information than just the data corresponding to its current value. The additional information relates to the history of the signal and is stored in predefined attributes. The predefined attributes that are particular to signals and not to variables are: STABLE, QUIET, TRANSACTION, EVENT, ACTIVE, LAST_EVENT, LAST_ACTIVE, and LAST_VALUE. Their detailed description is provided in the VHDL LRM [12] as amended by the document issued by the standard maintenance committee, VASG, [11].

3. Processes are sensitive to signals and not to variables. Namely, a process is executed when one of the signals on its sensitivity list changes value.

4. Signals can be global to more than one process, whereas variables are visible only in the process where they are defined.

It is important to understand the difference between a variable and a signal. It is the rich features of signal processing that are crucial to building the switch algorithm.

1.2.3   Concurrent vs. Sequential Statements

Concurrent statements in VHDL can refer only to signals. Therefore, the order of the execution of concurrent statements is not important since the execution of one statement does not affect the execution of another statement during the same simulation cycle. All assignments to signals are scheduled at minimum one simulation cycle later. Therefore, all concurrent statements get access to the same value of a signal at a given simulation cycle. Concurrent statements in VHDL are equivalent to processes.

One concurrent statement, the VHDL process construct, allows the definition of variables, and the use of sequential statements that refer to variables. It is only internal to a process that variables and sequential code may be used.

As a special characteristic of sequential statements, note that all sequential signal assignments made to the same signal use the same driver, whereas each concurrent signal assignment uses its own driver. It follows that a second sequential signal assignment within one process may preempt any previously scheduled assignment.

1.2.4   User-defined Packages

The concept of a package in VHDL is a simplified version of the one in Ada. VHDL packages may not be generic like Ada packages. VHDL package specifications may contain global signals. This is an extension to the concept of package in Ada.

A VHDL package has a specification and, optionally, a body. Whatever is defined in the body is not visible in any other compilation unit. The items defined in the package specification are visible in any compilation unit that uses the given package, including the package body. The basic idea is hiding information: only what needs to be seen should be made available. What needs to be defined in order to support the intended functionality, but is not directly used anywhere else, is defined in the package body. This concept has been borrowed from hardware where only few of the signals in a circuit are connected to the pins of a chip or of a PC board.

The global signals that may be defined in package specifications may be directly referenced only from within VHDL processes. Note that variables cannot be shared between two processes, since variables cannot be defined in architectures, nor in packages.

Section 1.3 presents in detail a VHDL package that was developed to support switch-level modeling.

1.2.5   User-defined Types

This is a familiar feature for Ada or C programmers. For hardware designers it may not be, so the concept is introduced from their point of view.

A useful, albeit simplified, view of a VHDL type is that it is "a series of indications of how to make a bundle of wires out of smaller bundles and/or wires". There are two cases to consider:

1. If the bundles that make a larger bundle are of the same kind, the way to bundle them in a larger bundle is by using the array construct. We say that all the little bundles are elements of the array. Arrays can be of specified size, generically constrained, or unconstrained. This means that the actual number of sub-bundles can be specified at the time the actual simulator is built (i.e., elaboration time) or at the time the type is defined.

2. If the bundles that make a larger bundle are of different kinds (i.e., different sizes, serve different purposes, etc.) the way to bundle them is using the record construct.

A more complex view is to consider that bundles have a particular color. Bundles of the same color can be connected (wire to wire) together. Bundles of different colors (i.e., different types) need explicit type conversion if they are to be connected together.

The color of the bundle indicates more than just the way the wires should be bundled. It may indicate the resolution function to be used. In this case, the resolution function specifies what value each wire of the bundle should have given that a number of drivers are connected to the bundle.

VHDL types are also used for abstracting data types. The analogy with the hardware is that of a bundle of wires that are always used together. In this case all possible values of the bundle, obtained by enumerating all possible combinations of values of the wires, can be listed as names of an enumerated type. This represents a different level of abstraction. For example, the output of a two-bit counter O1 and O2 can be modeled either as two ports of type bit having four possible values ('0','0'), ('0','1'), ('1','0'), and ('1','1'), or as one port having four possible integer values 0, 1, 2, and 3.

1.2.6   Value-System and Resolution Functions

The question of what data to communicate between the distributed components leads to the design of a value-system, i.e. a precise definition of the values that can be communicated via ports.

The problem of the value-system is crucial for interpretative and distributed algorithms which rely more than compiled algorithms on the interpretation of data passed and which must communicate all the information via the values of the value-system (no access to global data).

Digital simulations use a limited set of values to model the state of electrical signals in a circuit. The values that are absolutely necessary are 0 and 1. In modeling circuits it is sometimes necessary to use larger sets of values. The set ? (or X) representing "all values", and the set * (or Z) representing "no value" complete the values 0 and 1.

A next step in modeling refinement is to take into account the strength of a given value. The cartesian product of all strengths with all values represents the set of all possible values of an electrical signal.

Some sets, such as ? and * are absolutely necessary for modeling. Other sets turn out to be useful in an attempt to maintain all the available information. The use of subsets of the possible values provides for less conservative models.

By including all possible value subsets along with all possible values in the set representing the value-system, it is sure that information can be preserved in all circumstances in the model and models can be written as non-conservative as possible.

However, the number of all subsets is $2^{**}N$, where N is number of values, $N = 2^*S$, where S is the number of strengths. Therefore, the total number of subsets is $2^{**}(2^*S)$, which can become very large when S is large.

The various approaches proposed in literature consider only some of the subsets, on the basis that they are sufficient to solve all modeling problems at hand.

Counter-examples can be produced to show that using just some of the total number of subsets produce conservative models. If these counter- examples are not encountered too frequently and if the conservativism introduced is acceptable, then the subset-system of values is considered acceptable.

Modeling the conflict of two outputs driving the same wire occurs often enough to make it worthwhile having a special language construct for this purpose. VHDL allows the user to define resolution functions that are invoked whenever there are new contributions to a signal. A resolution function accepts as parameters all simultaneous contributions to a signal and returns a single value.

For a simple value-system, such as (0, 1, X, Z), there is a need to consider several resolution functions (wired-or, wired-and, tri-state, etc.), each using different assumptions regarding the relative "strengths" of 0 and 1. In the case of complex value-systems that incorporate the concept of strength, a single resolution function is sufficient to represent all conflicts to be resolved. Because VHDL can tie the resolution function to the type of a signal, it is more desirable to utilize only a single resolution function.

The circuit in Figure 1.1 highlights the potential loss of information associated to distributed algorithms. There are two possible approaches.

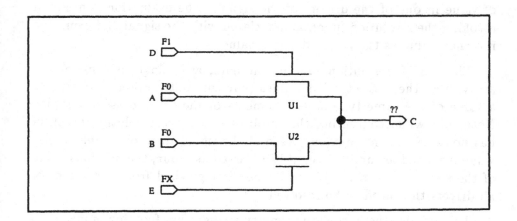

Figure 1.1: Handling of unknown

The first one, called conservative, is to say that since the value on the gate of U2 (supplied by port D) can be either 0 or 1, the value at the output (drain) of U2 may also be 0 or 1, and therefore is modeled as unknown. When resolved with the value zero contributed by U1, the conservative approach results in an unknown value.

A second approach preserves more information. The output of U2 is not unknown. It can take either the value F0 or the value ZX (i.e., either a strong zero or a much weaker unknown). Based on this information, the resolution function resolves the conflict between the outputs of U1 and U2 to F0, because F0 produced by U1 is stronger than ZX that might be produced by U2.

Therefore, by having sets of values in the value-system and by using such sets of values in order to preserve information, more useful simulations can be achieved.

As mentioned earlier VHDL allows the user to define both the type of the wire-values and the resolution function. There could be different resolution functions throughout the model and potentially each signal could have its own resolution function.

The evaluation of a signal is performed each time there is a change of value in one of the drivers of the signal. The evaluation consists of invoking the resolution function associated with the signal and using the returned value as the resolved signal value.

The VHDL resolution function accepts, as a single parameter, an array of all the current driver-values (current contributions). It returns a value of the same type as the elements of the array passed as input. Note that while computing the resolved value, the resolution function has no access to the previous value of the signal to resolve, unless that value is passed as input. Even if it is passed as input, the previous value of the signal to be resolved cannot be distinguished from the values of all drivers that need to be resolved.

This mimics the hardware property a resolution function is meant to model, but severely limits its modeling usefulness.

A driver may stop driving a signal if it corresponds to a guarded assignment for which the guard has become false.

VHDL provides the ability to control what happens when there are no more drivers driving a signal. The user can choose the kind of signal: a register or bus signal. In the case of a register signal, when the last driver disappears, the signal is not evaluated. In the case of the bus signal the resolution function is called with an empty input array.

Signals that do not have a resolution function associated with them can only be driven by a single driver.

1.2.7   Properties of Resolution Functions

Resolution functions may have some important properties. Understanding these properties is important when the user has the ability to create new resolution functions, as it is the case in VHDL.

The main properties of resolution functions are:

Lattice:

The values of the value-system can be partially ordered by strength. If the resolution function returns the smallest value that is larger than each of the values it is resolving, the value-system is said to be organized as a lattice. This is an important property for switch-level descriptions.

Associativity:

A resolution function is said to be associative if the resolved value is not affected by the replacement of any subset of driver-values by their resolved value.

Driver Order Invariance:

A resolution function is said to be driver order invariant if the resolved value does not depend on the order of the driver-values to be resolved.

Reflexivity:

A resolution function is said to be reflexive if the resolved value of any value is that same value.

The package presented in the next section introduces a resolution function that has all the properties mentioned above.

1.3   Switch-Level Package

Switch-level capabilities in VHDL are required to be supported by a value-system and the operations defined on it. This support is presented in the form of a VHDL package.

This section presents the 46-value system and its public domain associated package called std_logic[2].

The 46-value system has been designed by the present author and Andy Tsay [10]. The 46-value system is also described in [4].

The package std_logic was developed by several contributors, and is available from Vantage Analysis Systems Inc. for distribution costs.

The value-system and coresponding package are briefly introduced in the next sections. More details can be found in [4] or in the electronic disk that accompanies this book free of charge. The disk contains all the VHDL code necessary to run all examples in this chapter on any full-VHDL simulator.

---

[2]Copyright ©1988 Vantage Analysis Systems, Inc.

1.3.1   46-value System

The 46-value System is based on interval notation ([6] and [5]), in that it contains references to the subsets of possible values as defined with the interval notation.

The intervals considered are based on 9 ordered values: F0, R0, W0, Z0, D, Z1, W1, R1, F1. The system is extended with the uninitialized value, U.

As shown above, the value-system is based on two values (0 and 1), four strengths (F,R,W,Z), a disconnect value (D) that is so weak that it is neither 0 nor 1, and a value U that is similar to "all values" or FX. U provides the information that the corresponding signal has not been computed explicitly by a VHDL process during simulation. A signal having the value U must have had either no drivers or at least one driver with the value U, during each evaluation of that signal since the beginning of the simulation.

The value U is the strongest value in the simulation system. A conflict between U and any other value is resolved to U.

The value D represents the value of an undriven wire that has such a small capacitance that it cannot memorize whether it has been driven by a value 0 or 1 before being disconnected. Having a signal driven by the value D is equivalent to having the signal not driven at all. Being able to refer to D, makes the value system more general and provides a better interface for other hardware description languages. Also, refering to sets of values that include D proves to be useful.

The value-system begins with U and continues with values ordered based on the following rules, applied hierarchically:

1. the interval with the weakest of the stronger interval boundary first

2. the smallest interval first

3. the interval having 0 associated to the stronger boundary first

Whenever the same character appears twice in a row in the interval notation it is represented only once. Thus, FF0 becomes F0, RRX be-

comes RX, etc. Based on these rules, the following table presents the
names used to reference each interval:

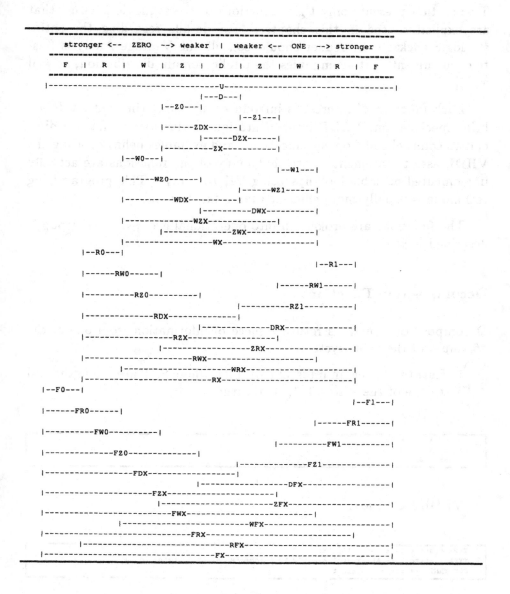

Figure 1.2: 46-Value System

1.3.2   Functional Support for the 46-value System

This section presents only those functions of the std_logic package that are explicitly used in this chapter. For details regarding the entire std_logic package the reader should consult the Vantage Analysis Systems document that acompanies the public domain distribution, as well as [4]

Each function of interest is introduced by name (in bold), with its brief specification, VHDL implementation for functions that are difficult to comprehend from specifications, and examples behavior using the VHDL assert statement. Note that many of the functions are actually implemented as table look-ups using VHDL arrays. This programming technique is usually more efficient than a function call.

The functions are broken up into operational groups. Each group is described next.

Decomposition Functions

Decomposition functions return a piece of information from one of the 46 values of the value-system.

1. **f_state:** Given an interval (SA,TB), where S, T are strengths and A, B are one of the states 0, 1, X, return:

```
'0' if A = B = 0,
'1' if A = B = 1,
'X' otherwise.
```

VHDL Definition:

```
TYPE t_state IS ( '0', '1', 'X' );
TYPE f_state_t IS ARRAY (t_logic'LOW TO t_logic'HIGH) OF t_state;
CONSTANT f_state : f_state_t := ( ... );
```

Example of use:

```
assert (f_state(U) = 'X');
assert (f_state(D) = 'X');
assert (f_state(RWX) = 'X');
assert (f_state(RW0) = '0');
```

2. **f_strength:** Given a value, return the maximum strength of all values in the interval. Specifically, given an interval (SA,TB), it returns:

```
 max(S,T),
'F' for interval U,
'U' for interval D.
```

VHDL Definition:

```
TYPE t_strength IS ( 'U', 'Z', 'W', 'R', 'F' );
TYPE f_strength_t IS ARRAY (t_logic'LOW TO t_logic'HIGH) OF t_strength;
CONSTANT f_strength : f_strength_t := ( ... );
```

Example of use:

```
assert (f_strength(U) = 'F');
assert (f_strength(D) = 'U');
assert (f_strength(RWX) = 'R');
assert (f_strength(WRX) = 'R');
```

3. **f_ceil:** Given a value, return a value of the same state, but with a strength not exceeding a value specified in the parameter "size".

VHDL Definition:

```
TYPE t_strength IS ( 'U', 'Z', 'W', 'R', 'F' );
TYPE f_strength_t IS ARRAY (t_logic'LOW TO t_logic'HIGH) OF t_strength;
FUNCTION f_ceil ( s: IN t_logic; size: IN t_strength) RETURN t_logic;
```

Example of use:

```
assert (f_ceil(F0, R) = R0);
assert (f_ceil(RWX, Z) = ZX);
```

Composition Functions

Composition function accept as parameters items of information and build one of the 46-values.

1. **f_logic:** Given a state and a max strength, return a value. Specifically, it returns:

```
        SA for strength S and state A.

        D for STRENGTH 'U',
```

VHDL Definition:

```
TYPE f_str IS ARRAY (t_strength'LOW TO t_strength'HIGH) OF t_logic;
TYPE f_log_con_t IS ARRAY (t_state'LOW TO t_state'HIGH) OF f_str;
CONSTANT f_logic : f_log_con_t := ( ... );
```

Example of use:

```
assert (f_logic('0')('U') = D);
assert (f_logic('0')('Z') = 20);
assert (f_logic('X')('W') = WX);
assert (f_logic('1')('R') = R1);
assert (f_logic('0')('F') = F0);
```

Comparison Functions

Comparison functions provide the capability to compare any two values of either type t_logic or t_state. The functions first convert any argument of t_logic type to t_state type, and then apply the comparison functions on t_state type. All the VHDL predefined relational operators are overloaded and supported with this functionality.

Tri-State Outputs

1. **f_convz:** Given a logic value, it returns an appropriate high-impedance tri-state logic value. Specifically, it returns:

```
f_convz(D) - D,
f_convz(U) - ZX,
f_convz(s0) - Z0 for any strength s,
f_convz(s1) - Z1 for any strength s,
f_convz(sX) - ZX for any strength s.
```

VHDL Definition:

```
TYPE f_logic_T IS ARRAY (t_logic'low to t_logic'high) OF t_logic;
CONSTANT f_convz : f_logic_t := ( ... );
```

Example of use:

```
assert (f_convz(RX) - ZX);
assert (f_convz(F1) - Z1);
assert (f_convz(Z0) - Z0);
```

2. **f_uxfr:** It models a uni-directional pass-transistor. It calculates the value of drain ('drn') as a function of gate ('g'), source ('src'), and drain ('drn'). It handles unknown values on 'g' without loss of information. Specifically:

```
IF g is on THEN
   the output to drn is src
ELSIF g is off THEN
   the output to drn is f_convz(drn)
ELSE
   the output to Drn is the interval between src and f_convz(drn)
END IF;
```

VHDL Definition:

```
TYPE f_logic_T IS ARRAY (t_logic'low to t_logic'high) OF t_logic;
FUNCTION f_uxfr(g, src, drn: IN t_logic) RETURN t_logic;
```

Example of use:

```
assert(f_uxfr(F1, F0, F1) - FX);
assert(f_uxfr(FX, F0, F1) - FZX);
```

Resolution Function

Given a vector of values, it returns the resolved value. Since the vector contains sets of values as its elements, all possibilities are considered by applying the resolution function to the cartesian product of all the sets supplied as input to the resolution function. The resolution function returns the narrowest interval containing the resolved values corresponding to the cartesian product of all sets of values provided as elements of the input vector.

VHDL Definition:

```
FUNCTION f_logic_bus( s : t_logic_vector ) RETURN t_logic;
SUBTYPE t_wlogic IS f_logic_bus t_logic;
```

Example of use:

```
1) s1 <- Z0; s1 <- Z1;
   assert ( s1 - ZX );
2) s1 <- WZ0; s1 <- WO;
   assert ( s1 - WO );
3) s1 <- WZ0; s1 <- WO; s1 <- FDX;
   assert ( s1 - FWO );
```

Note that the resolution function is called implicitly because there are several concurrent assignments to the same resolved signal.

1.4 Distributed Algorithm for Pass-transistor

The 46-value package is sufficient to model a bi-directional MOS switch logically. Such a model is now introduced. Earlier versions of this algorithm are detailed in [4] and [9].

1.4.1 Modeling Assumptions

Before presenting the algorithm, we need to understand the modeling assumptions that are made, both in terms of limitations and in terms of capabilities.

Commutation:

A switch can be either on or off. A non-zero delay must be specified for the commutation time of the switch. This ensures that a change on the gate cannot occur during the local relaxation procedure. This is necessary due to limitations of the VHDL implementation.

Zero delay between source and drain:

This corresponds to source and drain capacitances that are low compared to the other capacitances (high) and resistors (low).

Superposition:

Many weak drivers may not override a stronger one.

Driving strength:

A switch may not drive stronger than a certain user-specified limit. This allows one to model complex networks, such as the RAM cell presented in section 1.6.1.

Interval-Notation Value-System:

The use of an interval-notation - based value-system provides for useful simulation results by limiting the unknown predictions. The example in figure 1.1 illustrates this capability.

Power supply:

Signals that can be driven with the strength 'F' should be driven only with that strength. This limitation is due to an optimization that is based on this assumption. Pass-transistors will not attempt to modify a signal with the strength 'F'. This speeds up the simulation of circuits that have many transistors connected to the same two signals: power and/or ground. Many loops are disconnected if no activity is propagated through power supply, and the simulation completes much faster.

1.4.2 Overview of Algorithm

The algorithm for simulating networks of bi-directional switches is distributed in the sense that it consists of the co-operation of switch instances during simulation. All switch instances are governed by the same algorithm, referred to in this chapter as the distributed algorithm for bi-directional pass transistors.

Note that each switch instance communicates with the other instances solely via the values placed on the three signals to which it is connected: gate, source, drain. Also, at the end of each time point, the value placed on each signal contains only information regarding the predicted value for that signal and not extra information, such as direction of the signal flow, etc.

The idea behind the distributed algorithm is to disconnect the source and drain from the network by driving them with the weakest strength that preserves the value (i.e., Z), in order to find out what the contribution of the network to the values of those signals is. Based on that information, the switch reconnects by driving both the source and the drain with the resolved values of the source and the drain as contributed by the network.

The distributed algorithm of the bi-directional switch consists of watching the inputs and recomputing the outputs by disconnecting and then repeatedly reconnecting the switch to the network.

The state information of the switch consists of the value of the internal gate (gd), the values with which the ports Src and Drn were driven up to the current time, the values with which Src and Drn must be driven next, the initialization flag, and information regarding whether the switch is ready to disconnect or it is reconnecting.

Besides initialization, the two operations performed by the switch are disconnection and reconnection.

1. **Disconnection** - Whenever there is activity on any signal to which an on-transistor is connected, disconnect the on-transistor from the source and drain connections. Whenever there is activity on any signal to which an off-transistor is connected, disconnect the off-transistor only from the connection(s) (source and/or drain) where the activity occurred. Disconnecting a transistor from a signal is achieved by having the transistor drive the given signal with the same state as on the net but with the strength Z.

   Note:

(a) A signal that is not driven preserves its state (zero or one), but changes its strength to 'Z'.

(b) The disconnection of a transistor represents an activity on the signal.

2. **Reconnection** - During the delta that follows the disconnection, drive the disconnected ports with the resolved value of the drain and source ports if the transistor is on and with the state found after disconnecting the source or drain and 'Z' strength if the transistor is off.

In summary, for any particular cycle, a transistor could be a) untouched by the disconnection "wave" and ready to disconnect as soon as either there is an event on gate or there is a transaction on Src or Drn or b) disconnected and reconnecting either actively (if there is activity on Drn or Src) or passively (if there is no activity on its Drn and Src ports).

The algorithm ceases the local activity as soon as a match between the old driving value and the new computed one is detected. This produces a local activity pause. If all transistors are in local activity pause, a global activity pause is reached.

Switches may not be turned back into ready to disconnect state during any of the local activity pauses. One solution to enforce this constraint is to use the minimum VHDL time unit (1 fs) and restrict a switch from going back to ready to disconnect for 1 fs from the time it was first disconnected. No time restrictions are further necessary since all computational activity take only delta cycles (simulation cycles for which the time is not advanced). As soon as the global activity ceases the time is advanced to the next fs and the transistors are ready to be disconnected again.

Another solution to the same problem is to save the simulated time (using the pre-defined function NOW) associated to a disconnection and to consider the transistor in the state wakeup = true, if the current time is equal to the saved time.

Note that events that are 1 fs apart are handled correctly by the relaxation algorithm, since a disconnection/reconnection is done in one time point. Hence, any combination of delays are handled correctly, including real and/or mutually prime values.

The VHDL code corresponding to the algorithm is presented in section 1.5.

1.4.3 Completion of Distributed Algorithm

The following properties of the algorithm are sufficient to ensure that there are a finite number of disconnection and reconnection steps at each time point. Also, the value of all signals on an open path at the end of each time point is the resolved value of all signal values present on the open path at the beginning of the time point.

Property 1: The non-zero, natural, delay introduced on the gates interrupts immediate (delta delay) feedback via gates. Corollary 1.1: during the reconnection step, the transistor does not change its on/off state.

Property 2: A switch is never disconnected twice. Indeed, the implementation presented here has this property. After disconnection the state becomes wakeup = true, and no other disconnection can occur for the same transistor.

Property 3: All switches on an open path are disconnected, iff any of them disconnects. Indeed the implementation has this property: all on-transistors that are contiguous to a transistor that disconnects will disconnect because each transistor is sensitive to the disconnection of its neighbors. By induction the entire path will disconnect. Note that the off transistors are already disconnected, as they drive with strength Z. They may just change their driving state.

Property 4: The resolved value of two intervals is the smallest interval containing all possible resolved values from both intervals. Indeed the resolution function of the std_logic package has this property.

Property 5: Switches do not increase the strengths of the values they pass. Indeed the implementation has this property. Each switch drives with at most the strength of the resolved values of the two ports Src, and Drn. The resolution function, according to property 4, is such that the resolved value is an interval with bounds that are contained in one of the values to be resolved.

Property 6: The resolution function is associative. That is $f(f(a,b),c)$ = $f(a,f(b,c))$, where f is the resolution function in the std_logic package and a,b, and c can be any value of the 46-values of the value system under consideration. Indeed, this is the case for the resolution function used in the implementation.

Property 7: The resolution function returns the same value for any permutation of its arguments.

Property 8: All disconnected switches begin reconnecting. Indeed this is the case for the implementation under consideration since the switch is sensitive to transactions on Drn and Src. By disconnecting, each transistor starts its own reconnection.

Theorem 1: The disconnection phase affects all switches on an open path where one switch is affected. Such a disconnection always completes in a finite number of simulation cycles.

Proof: Let us consider a simplified model consisting of n signals $s_i, i = 1...n$ connected by m bi-directional pass transistors. The gate connections are not significant for this step since internal gates (gates after delay) are active only in the first cycle of a time point (Prop 1).

All transistors on an open path (including the off ones contiguous to the path) disconnect (Prop 3). Each switch-process stays in the state wakeup = false until there is some activity on src, drn, or an event on gd. Starting with switches connected to active signals, each open or possibly[3] open switch is disconnected. This process is not repeated twice for any switch (Prop 2), since for any other cycle the signal wakeup is true. All switches on a path are disconnected and a path does not include a closed switch. The finiteness of each path, and the fact that disconnection is not repeated for any switch (Prop 2) ensure the termination of disconnection.

Theorem 2: The reconnection phase affects all switches and completes in a finite number of simulation cycles.

Proof: All nodes touched by disconnected transistors in the recon-

---

[3] 'X' state on the gate

nection state may potentially change their value during reconnection, because they all begin the reconnection (Prop 8).

Let us label such nodes as red nodes and let n be their total number. A source node is a node that will not change its value during the reconnection stage. Let us denote by primary the red nodes having the strengths of their value equal to the strength returned by the resolution function applied to all red nodes. Note that the strength of an interval value is the strongest strength of the two boundaries.

Let k be the number of primary nodes. There are two cases to consider:

i) All strongest values are equal to each other. Because of Prop 4 it follows that during the reconnection stage all primary nodes will not change their values and can be considered sources.

ii) Primary nodes have different values. There are two cases to consider:

a) Some primary nodes may be isolated from the other because the switches on all paths between them decrease the strengths of the values they pass. Such nodes will not change their values during reconnection because of Prop 5, and can be considered sources.

b) Some primary nodes are connected to the other primary nodes by paths that do not decrease the strengths of the values they pass. After enough number of simulation cycles during which the reconnection is performed all primary nodes that are connected will have as their value the resolved value of their original values. This is so because the number of switches is finite and because the resolution function is associative and of permutable parameters (Prop 6 and Prop 7). Once the primary nodes have the resolved value of all primary nodes, they will not change their value during the current reconnection stage. Therefore they can be considered sources.

We have shown so far that after a finite number of cycles all primary nodes can be considered sources and the remaining n - k red nodes (where k > 0) represent a new set of red nodes that may change the values during the reconnection activity. It follows by induction that after a finite number of cycles all originally red nodes become sources. This is the termination condition for the reconnection activity.

Theorem 3: The number of cycles necessary to resolve n contiguous on-transistors (island of n transistors) is not higher than 2n.

Proof: The reconnection activity is done in parallel for the whole circuit, and one cycle behind the disconnection. Hence, the overall number of cycles necessary to decide the values for one time point is equal to or smaller than twice the number of transistors in the biggest open island and it is not larger than 2 m + 1. Hence, the number of VHDL cycles is linear with respect to the number of transistors in the circuit.

Theorem 4 : for a network composed only of switches, the values of all signals on a pass of on-transistors at the end of each time point is the minimum value larger than any of the values of the signals under consideration at the beginning of the same time point.

Proof: use the same proof as for Theorem 2. At completion all signals have the resolved value of all red signals. By induction, all signals have the resolved values of the first set of red signals, namely the resolved value of the values ahving the maximum strength on the given open path. This value is equal to the resolved value of all signals on the open path according to property 4.

In effect the proof follows from the lattice property of the resolution function over the given value-system.

1.5   VHDL Implementation of Distributed Algorithm

Now that the algorithm is understood, its implementation in VHDL can be introduced.

The local relaxation algorithm is implemented in VHDL using the following features: process (sensitive to the internal gate and to the source and drain), user-defined resolution function (supporting the lattice organization of the value-system), predefined attributes (transaction that senses any activity on the source and/or drain), user-defined types (for the 46 values of the value-system), internal signals (internal gate: gd, and

wakeup that is used to test that there are no more transactions for the
current time point), static variables within processes (see the comments
within the process).

The actual VHDL description of a bi-directional pass transistor is
presented below. It depends heavily on the functions of the std_logic
package.

```
USE work.STD_LOGIC.ALL;
entity bxfr is
    generic (gdelay : time := 3 ps; MaxStrength : t_strength := 'R');
        -- gdelay       : delay on the gate G
        -- MaxStrength : max output strength on Src and Drn. Used to model resistance.
    port( g: in t_wlogic; src,drn: inout t_wlogic);
        -- Src and Drn are symmetric.
end bxfr;
--
architecture bxfr_behavior of bxfr is
        signal wakeup : boolean := false;
        signal gd : t_wlogic;
begin
    gd <= g after gdelay;
    process(gd, src'transaction, drn'transaction)
        variable psrc, pdrn : t_logic;    -- to store old  contributions
        variable vsrc, vdrn : t_logic;    -- to store next contributions
        variable init : boolean := true;  -- to initialize src and drn at powerup
    begin                                 ----------------------
        if not wakeup then        ----- disconnect -----
            if init then                  ----------------------
                src <= D; drn <= D; init := false;
            elsif (gd'event or (f_state(gd) /= '0')) then
                wakeup <= true;
                psrc := f_convz(src); pdrn := f_convz(drn);
                src <= psrc; drn <= pdrn;
            else -- no event on gate, and gate is off
                if src'active then  -- evaluate src
                    vsrc := f_convz(src);
                    if psrc /= vsrc then  -- update when change
                        psrc := vsrc;
                        if f_strength(src) /= 'F' then
                            src <= vsrc;
                        end if;
                    end if;
                end if;
                if drn'active then  -- evaluate drn
                    vdrn := f_convz(drn);
                    if pdrn /= vdrn then  -- update when change
                        pdrn := vdrn;
                        if f_strength(drn) /= 'F' then
                            drn <= vdrn;
                        end if;
                    end if;
                end if;
            end if;
        else                          ----------------------
            case f_state(gd) is   ---- reconnect -----
                when '0' =>           ----------------------
```

```
                                -- f_convz computes tristate value
                vsrc := f_convz(src);
                vdrn := f_convz(drn);

        when '1' =>         -- compute brf of src and drn
                           -- f_ceil computes the resistance
                           -- f_busres computes the resolved value
                vsrc := f_ceil(f_busres(src)(drn),MaxStrength);
                vdrn := f_ceil(f_busres(src)(drn),MaxStrength);

        when 'X' =>        -- compute range between tristate and brf of src and drn
                          -- f_uxfr is the function for unidirectional transistor
                          -- use f_uxfr to compute the range between tristate
                          --     and resolved value
                vsrc := f_ceil(f_uxfr(FX,f_busres(src)(drn),f_convz(src)),MaxStrength);
                vdrn := f_ceil(f_uxfr(FX,f_busres(src)(drn),f_convz(drn)),MaxStrength);
        end case;
        if   (psrc = vsrc) and (pdrn = vdrn) then -- wakeup in 1 fs, if nothing changes
                wakeup <= false after 1 fs;
        else
            if psrc /= vsrc then  -- update when change
                psrc := vsrc;
                if f_strength(src) /= 'F' then
                    src <= vsrc;
                end if;
            end if;
            if pdrn /= vdrn then  -- update when change
                pdrn := vdrn;
                if f_strength(drn) /= 'F' then
                    drn <= vdrn;
                end if;
            end if;
        end if;
    end if;
end process;
end bxfr_behavior;
```

The bi-directional switch is described as an entity with two generic parameters (Gdelay, and Maxstrength), and three external ports G, SRC, and DRN corresponding to gate, source, and drain. Gdelay represents the delay on the gate and models the speed of commutation. MaxStrength represents the maximum strength with which the switch may drive a signal connected at its drain or source and models the voltage drop between source and drain. Its use is exemplified in the example presented in section 1.6.1.

The architecture represents the execution code of the model. It consists of two concurrent statements: a delayed assignment and a process.

The delayed assignment of G to the internal signal gd ensures that a transaction on gd can occur only during the first simulation cycle of a time point.

The process is sensitive to the events on gd, SRC'transaction, and DRN'transaction. When gd changes its value, the values on SRC and DRN need to be recomputed in response to the commutation of the switch.

The attribute 'transaction is used to propagate the disconnection even though the signal values may not change. Note that S'transaction is a signal that has events whenever S has transactions[4]. It is important to use S'transaction because processes in VHDL are sensitive only to events on signals, and the model of the bi-directional switch must react to any transaction SRC and DRN in order to propagate the disconnection along a path of open switches.

Variables are used to store the values previously driven by the process (psrc and pdrn), as well as the values that have to be driven next (vsrc and vdrn). The same values are not assigned again while reconnecting. This ensures the completion of the process activations for a particular time point.

An internal wakeup signal is used to model the connecting and the disconnecting states. When the wakeup value is false, the process is in a disconnecting state. When the wakeup value is true, the process is in a connecting state, and will be changed to false after 1 fs inertial delay. The 1 fs inertial delay is to guarantee that there are no events on the sensitivity list of the process, when the wakeup value changes from true to false.

In the disconnecting state, the pass transistors remove their driving effects. In the reconnecting state the pass transistors "pass", in both directions, the resolved values of the driving sources.

Note that pass transistors do not "drive" signals, they just "pass" the values from the driving sources. The driving sources come from power, ground, or external gates.

---

[4]Any change in the value of a driver of a signal S represents a transaction on S. A transaction on S may not result in an event on S, because the resolved value of all drivers of S may produce a value equal to the current value of S.

1.6   Examples of Switch-level Networks

Given that the model for the bi-directional switch now exists, networks of the model can be built.

This section presents the VHDL source code for switch-level networks. The complete code is available on the accompanying disk. The code uses the std_logic package.

1.6.1   One-bit RAM

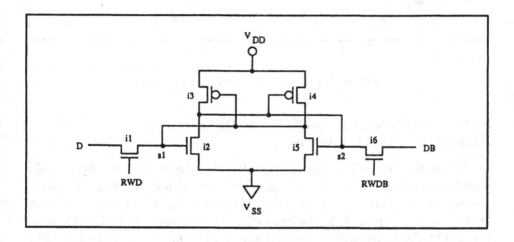

Figure 1.3: One-bit RAM

Figures 1.3 and 1.4 present a one-bit RAM consisting of six transistors and the corresponding simulation results. Note that the legend describes the graphical encoding of the strengths. In the one-bit RAM circuit, the two transistors connected to the outside world have a strength of 'R' and the other transistors have a strength of 'W'.

The inputs RWD and RWDB represent the read/write controls and they open and close the connection to the outside world. The input D is driven to resistive values (R0 or R1), or to tri-state values (Z0 and Z1). Note from the simulation results that when RWD/RWDB are R0, the

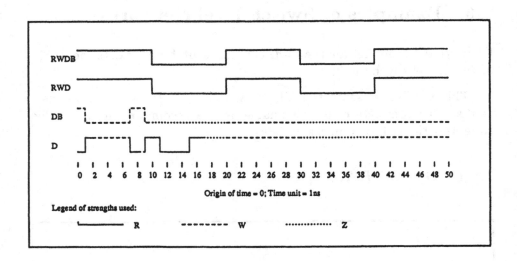

Figure 1.4: One-bit RAM - Waveform

RAM is disconnected and both D and DB are driven with strength 'Z' (keeping the existing state 0 or 1).

When RWD/RWDB are R1, the RAM is connected to the outside world and D and DB are driven from the RAM with a strength 'W' which comes from the internal transistors. If the outside world contributes to D/DB with a strength Z the operation corresponds to a READ since the RAM data overwrites the outside world and if the outside world contributes with strength 'R' the operation corresponds to a WRITE since the data from the outside world overwrites the data of the RAM, as it is shown in the transitions on DB at times 1ns, 7ns, and 9ns.

1.6.2   Memory Cell based on two Inverters

Figure 1.5 presents a memory cell that is made out of five switches labeled U1 through U5, which are described below:

U1 - The uni-directional switch U1 drives zeroes with the strength F (highest strength) and ones with the strength R (second highest strength). When its gate is driven with a zero its contribution to the value of signal

C is ZX, Z0, or Z1 depending on the last value with higher strength that was present on the signal C.

U2 - The uni-directional switch U2 is used for reading from the cell.

U3 - This is the only bi-directional switch in this circuit. The dataflow is left to right for write operations and right to left for read operations.

U4, U5 - constitute the actual memory of the cell.

The inverter model is built of networked switches similar to the NAND in figure 1.6. The code for the inverter and for the memory cell is included on the accompanying disk.

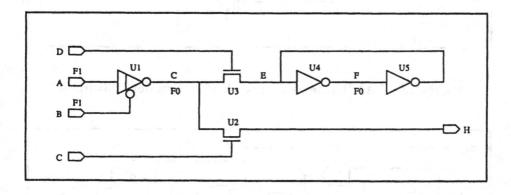

Figure 1.5: Memory Cell based on two Inverters

1.6.3   1,2,4 and 6-bit adders

The schematics of a one-bit adder is presented in Figure 1.7. The NAND gates used to implement the adder have the schematics presented in Figure 1.6. Note that i1 and i2 are pfet-transistors while i3 and i4 are nfet-transistors. The pfet and nfet transitor-models are based on the general bi-directional pass-transistor model presented in section 1.5.

The VHDL code corresponding to a NAND gate and the associated transistor definitions is presented below:

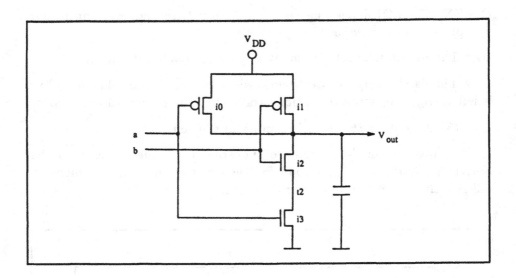

Figure 1.6: NAND gate as a network of bi-directional Switches

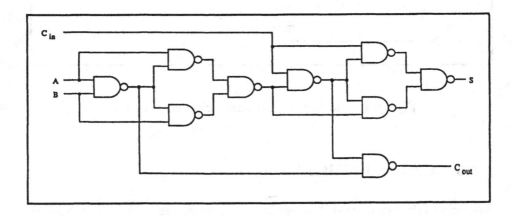

Figure 1.7: One-bit adder as a network of NAND gates

```
---------------------------------
-- nfet model used in Nand gate
---------------------------------
USE work.STD_LOGIC.ALL;
```

```
USE work.ALL;
entity nfet is
    generic (gdelay : time := 3 ps; MaxStrength : t_strength := 'R');
    port( g: in t_wlogic; src,drn: inout t_wlogic);
end nfet;
architecture nfet_behavior of nfet is
    component bxfr_type
        generic (gdelay : time := 3 ps; MaxStrength : t_strength := 'R');
        port( g: in t_wlogic; src,drn: inout t_wlogic);
    end component;
begin
    i1: bxfr_type generic map (gdelay, MaxStrength)
                  port map( g, src, drn );
end nfet_behavior;

--------------------------------
-- pfet model used in Nand gate
--------------------------------
USE work.STD_LOGIC.ALL;
USE work.ALL;
entity pfet is
    generic (gdelay : time := 3 ps; MaxStrength : t_strength := 'R');
    port( g:        in t_wlogic; src,drn: inout t_wlogic);
end pfet;
architecture pfet_behavior of pfet is
    component bxfr_type
        generic (gdelay : time := 3 ps; MaxStrength : t_strength := 'R');
        port( g: in t_wlogic; src,drn: inout t_wlogic);
    end component;
    signal tg : t_wlogic;
begin
    tg <= f_logic(f_not(f_state(g)))('R');
    i1: bxfr_type generic map (gdelay, MaxStrength)
                  port map( tg, src, drn );
end pfet_behavior;

----------------------------
-- nand model used in adder
----------------------------

USE work.STD_LOGIC.ALL;
USE work.ALL;
entity nand2 is
    port (a,b: in t_wlogic; z: inout t_wlogic);
end nand2;
architecture nand2_xfr of nand2 is
    component nfet_type
        generic (gdelay : time := 3ps);
        port( g: in t_wlogic; src,drn: inout t_wlogic);
    end component;
    component pfet_type
        generic (gdelay : time := 3ps);
        port( g: in t_wlogic; src,drn: inout t_wlogic);
    end component;
    signal t2  : t_wlogic;
    signal vdd : t_wlogic := F1;
    signal gnd : t_wlogic := F0;
begin
    i0: pfet_type port map( a, vdd, out );
    i1: pfet_type port map( b, vdd, out );
    i2: nfet_type port map( a, out, t2 );
    i3: nfet_type port map( b, t2, gnd );
```

```
end nand2_xfr;
```

One can see from this example how complex models can be quickly
built from the basic bi-directional transfer gate. Once a basic library
of switch-level logic elements is built utilizing the basic elements. Also,
library elements consisting of network of transistors can be recoded as
single processes for more efficient simulations.

1.6.4   Performance of VHDL Switch-level simulation

The table in Figure 1.8 provides data regading the performance of the
switch-level simulation in VHDL. The measured circuits are adders (1,2,4,
and 6 bit adders) made out of logic gates (9 per bit), where each logic
gate is made out of four transistors. The full models are supplied on the
accompanying disk.

The stimulus was supplied by a top-level VHDL module, and con-
tained 128 quadruplets that scanned all possible values eight times. The
VHDL simulator used for measurement was the VantageSpreadSheet
V1.1, on an Apollo DN3000.

```
Circuit | Nr. of switches | CPU time in sec.
--------------------------------------------------
Adder1  |       36        |        45
--------------------------------------------------
Adder2  |       72        |        64
--------------------------------------------------
Adder4  |      144        |        74
--------------------------------------------------
Adder6  |      216        |        80
```

Figure 1.8: Simulation execution time on an Apollo DN3000-6MB

Note that the CPU time depends on the activity in the circuit and
not on the number of transistors. Performances of more complicated
circuits have not been listed as it is difficult to describe the exact test
vectors used.

Although the performance is not at the level of a dedicated switch-
level simulator, the advantage of having true mixed mode switch and

behavioral simulations is very important. Also, this modeling technique for bi-directional pass-transistors can be used to specify hardwired primitives that perform much better. There will be an important portability benefit in having such primitives described in the standard language which is VHDL.

1.7   Future Research

There are many areas which can be focussed on to improve the quality and performance of the simulation. These are touched on briefly here.

1.7.1   Transistor-network Pattern Recognition

Pattern recognition of transistor networks (see [8], [7]) improve both the quality and the performance of switch-level simulations. The recognized pattern can be replaced with an equivalent description that models the behavior of the network more efficiently.

1.7.2   More Complex Value-System

Perhaps a more complex value-system (using as many as 4096 values) may be needed. The value-system used in [1] is based on two values and 255 strengths. It has the lattice property, but it does not benefit from the advantages brought by the interval notation. The results may be more conservative than when using an interval notation type of approach.

1.7.3   Hardwired Implementations

Efficient implementations of switch-level primitives are being introduced in VHDL simulation kernels (see [1]). Such implementations will become legitimate when pure VHDL descriptions for each primitive will be provided. At that time only, switch-level VHDL descriptions will be portable to any VHDL simulation platform.

1.7.4   Analog Models in VHDL

Analog VHDL descriptions are being developed (see [2], as well as the next three chapters). The value-system used is based on a real value accompanied by information regarding what the real value represents: current or voltage. It is the user's responsibility to connect pins together that are described by the same kind of values: either current or voltage. In the case of current the resolution function adds the contributing currents into a unique sink (currently only one sink is allowed).

More research is being put into this area. A special interest group of the VHDL Analysis and Standardization Group (VASG) has been organized to bring together people interested in the use of VHDL for mixed Analog-Digital simulations.

1.8   Conclusion

This chapter provided information on the programming capabilities available in VHDL. These capabilities were exercised in solving a challenging problem: modeling bi-directional switches in pure VHDL (i.e., without any extensions). Hopefully the reader is now able to apply the programming features of VHDL to solve other interesting problems.

1.9   References

[1] R.D. Acosta, S.P. Smith, Jeff Larson *Mixed-Mode Simulation of Compiled VHDL Programs*, ICCAD89.

[2] B.R. Stanisic, M. W. Brown *VHDL Modeling for Analog-Digital Hardware Designs*, ICCAD89.

[3] R.E. Bryant, *A Survey of Switch-Level Algorithms*, IEEE Design and Test of Computers, vol. 4, no. 4, pp.26-40, August 1987.

[4] D.R. Coelho, *The VHDL Handbook*, Kluwer Academic Publishers, Norwell, MA, 1989.

[5] P. Flake, P. Moorby, G. Musgrave, *An Algebra for Logic Strength Simulation*, 20th Design Automation Conference, IEEE 1983.

[6] P. Flake, P. Moorby, G. Musgrave, *Logic Simulation of Bidirectional Tri-state Gates*, Proc ICCC80, Port Chester, NY pp 594-600.

[7] D. Overhauser, I. Hajj, Y-F. Hsu *Automatic Mixed-Mode Timing Simulation*, ICCAD89.

[8] V.B. Rao, D.V. Overhauser, T.N. Trick, I.N. Hajj *Switch-Level Timing Simulation of MOS VLSI Circuits*, Kluwer Academic Publishers, Norwell, MA, 1989.

[9] A.G. Stanculescu, A.S. Tsay, A.N.D.Zamfirescu, D.L. Perry *Switch-Level VHDL Descriptions*, ICCAD89.

[10] A.G. Stanculescu, A.S. Tsay *A 46-State Value-System from Vantage Analysis Systems, Inc.*, VHDL User's Group VHDL Design Exchange Group meeting , February 89.

[11] The Computer Society of the IEEE's DATC Design Automation Standards Committee, *The Sense of the VASG*, Proceedings of the Second Fall User's Group Meeting, October 1989.

[12] IEEE, *VHDL Language Reference Manual*, IEEE std 1076-1987, N.Y., N.Y., March 1988.

[13] R. Lipsett, Carl Schaefer, Cary Ussery *VHDL: Hardware Description and Design*, Kluwer Academic Publishers, Norwell, MA, 1989.

Chapter 2

Modeling of
Transmission Line Effects
in Digital Circuits

Balsha R. Stanisic

International Business Machines
Rochester, Minnesota

This chapter is written to assist the applications engineer familiar with transmission line behavior in improving digital simulation by incorporating transmission line behavior into digital models. It describes an approach for simulating the combined digital logic-transmission line behavior and centers on the VHDL behavioral models used to simulate transmission line effects that enable simulation of off-chip environments such as multichip modules and cards.

2.1 INTRODUCTION

Traditionally, the "logic" behavior and transmission line behavior for a digital hardware design have been modeled and simulated separately. "Logic" behavior described by the (0, 1, X, etc.) level of abstraction can usually model on-chip digital networks with delay equations because interconnections typically exhibit negligible transmission line effects. However, at higher switching frequencies and greater interconnection lengths, transmission line effects become more significant making the "logic" level of abstraction less adequate for accurately modeling dynamic behavior.

To predict the dynamic behavior of off-chip and high speed on-chip digital networks exhibiting significant transmission line effects, the designer must typically model and simulate the transmission line behavior of select networks with a circuit simulator such as SPICE [7] or ASTAP (Advanced STatistical Analysis Program) [6]. Simulating this behavior verifies design aspects not checked by logic behavior such as incident switching, false switching, proper termination, and effective delay.

An environment capable of simulating the combined logic-transmission line behavior is more desirable in comparison to this separate simulation approach. The problems associated with partitioning and mapping are alleviated since the judgement of selecting which networks are "critical" or "complex" is removed. Also, the integrity of interconnection is maintained with one network description. More importantly, as the level of integration increases with multichip modules, the capability to prototype the design is reduced. Also, as operating frequencies increase with faster technologies and interconnection lengths increase with larger chips -- giving rise to transmission line effects on-chip -- improvements in the fundamental digital simulation model should be made.

VHDL provides an environment capable of simulating combined logic-transmission line behavior improving the fundamental digital simulation model. Primarily used to model digital "logic" behavior, VHDL also has the descriptive capability to model transmission line behavior for event-driven simulation. The behavior of lossless transmission lines interacting with drivers and receivers based on resistive loading effects can be easily modeled.

2.2 UNDERLYING CONCEPTS AND STRUCTURE

2.2.1 Superposition

The concept of superposition is fundamental to transmission line simulation. Transmission line simulation is based on traveling waves. The waves are referred to as incident, reflected and transmitted. Their aggregate effect at a point determines the voltage and current behavior at that point. Hence, summing or superimposing these waves is required to determine the voltage behavior at a node.

The summing is accomplished through the resolution function package of Figure 1.

```
PACKAGE summing_bus_res_function IS
--
-- Declare subtype of a real for bus resolution function that is
-- an unconstrained array.  The size of summing_val array equals the
-- number of signal assignment statements assigned to the common signal
-- that is currently calling the function.

   TYPE summing_val IS ARRAY ( integer RANGE<> ) OF real;
--
-- The function for bus resolution sums the currents or voltages at the
-- common signal (dot), hence, if one of the components is disabled its
-- signal assignment statement driving the signal should be assigned 0.0
-- in the behavior
--
   FUNCTION summing_brf_val ( data_array: summing_val )RETURN real;
--
   SUBTYPE brf_real IS summing_brf_val real;
--
END summing_bus_res_function;

PACKAGE BODY summing_bus_res_function IS
--

   FUNCTION summing_brf_val ( data_array: summing_val )RETURN real IS

      VARIABLE    i        : integer;
      VARIABLE    result   : real := 0.0;

      BEGIN
            FOR i IN data_array'RANGE
            LOOP
                result := result + data_array(i);
            END LOOP;

            RETURN result;
      END summing_brf_val;
--
END summing_bus_res_function;
```

Figure 1.   Basic Summing Bus Resolution Function

For multiple signal assignments to a common signal, this function simply sums the corresponding *brf_real* fields of the signal assignment drivers and applies it to the common signal. Hence, transmission line behavioral models implement superposition by using the signal types defined in the next section comprised of of these *brf_real* fields.

2.2.2 Identifying and Structuring Modeling Information

The first step in storing data is determining its units. The units for the modeling information form the consistent set of [volts], [milliamps], [kohms], [millimhos], [nanoseconds], [picofarads], and [microhenries]. This set is convenient for digital designs comprised of digital logic integrated circuits. The concept of an implicit consistent set of units is adopted from the ASTAP circuit simulator.

To illustrate its use, consider Ohm's law $V = RI$. Numerically expressed in exponential form, (VE00) = (RE3) * (IE-3), the values V, R, and I have their exponents expressed in the consistent set of units. Note this reduces to $V = RI$ eliminating the need for the exponent when the consistent set of units is used. This allows the models to operate solely on the numbers. Another example expressing Ohms's Law with conductances states $I = GV$. Again, (IE-3) = (GE-3) * (VE00) or $I = GV$. Hence, the units do not accompany any modeling information.

Several key items are required for transmission line simulation. Minimum and maximum resistance and conductance bounds are needed to approximate "open" and "short" circuit conditions. The signal mode (voltage or current) is necessary to identify the driver type. The total node voltage containing the result of the superimposed traveling waves is required to provide the primary simulation result. The wave components, namely the internally incident traveling voltage wave which works in conjunction with the total node voltage to determine subsequent voltage behavior, is essential. The transmission line branch current is needed to provide current magnitude information for each transmission line segment. Finally, the far-end voltage and current provide key information for calculating the "general driver" behavior of nonlinear drivers [1].

These items constitute the basic structures required for VHDL modeling of transmission lines. The transmission line data structure which supports these items for the models is shown in Figure 2.

The constants *g_min*, *g_max*, *r_min*, and *r_max* are used by the transmission line behaviors to bound the conductance and resistance specified by the user through VHDL generics. Also note that the units of these bounded values are in [Kohms] and [millimhos] which conforms to the consistent set of units [volts], [mA], [Kohms], etc., given earlier. The signal modes *v* or *i* identify the driver type (voltage or current) for the linear driver model (described later) so the correct output voltage is applied to a node.

---

[1]   For details refer to "general driver" section 2.3.4 and Figure 17.

```
USE work.summing_bus_res_function.all;
PACKAGE tline_data_structure IS
--
-- transmission line max and min load limits
--
    CONSTANT g_min : real :=       0.0001; -- 0.1  micro mho
    CONSTANT g_max : real :=   10000.0    ; -- 10.0 mhos

    CONSTANT r_min : real :=       0.0001; -- 0.1  ohms
    CONSTANT r_max : real :=   10000.0    ; -- 10.0 mega ohms
--
-- signals
--
    TYPE sig_mode IS (v,i);
--
-- transmission line node
--
    TYPE tline_node IS RECORD
      volt      :  brf_real;
      g0        :  brf_real;
      volt_fe   :  brf_real;
      cur_fe    :  brf_real;
    END RECORD;
    --
--
-- transmission line traveling signal
--
    TYPE tline_traveler IS RECORD
      volt:  brf_real;
    END RECORD;
    --
--
-- transmission line branch
--
    TYPE tline_branch IS RECORD
      cur :  brf_real;
    END RECORD;
    --
--
END tline_data_structure;
```

Figure 2.   Transmission Line Data Structure

The transmission line node, *tline_node*, is defined as a record consisting of four floating-point fields. The first, *volt*, stores the total voltage on the node. The second, *g0*, contains the total conductance of all elements connected to the node. The remaining fields, *volt_fe* and *cur_fe*, hold the far-end total voltage and branch current for nodes with a "general driver" connection. Similarly, *tline_traveler* and *tline_branch* each contain one field and store the transmission line internally incident voltage and branch current, respectively. Finally, as explained earlier, all fields are of type *brf_real* defined in the summing resolution function package of Figure 1. These two packages form the foundation for the general model structure shown in Figure 3 used to simulate transmission line behavior.

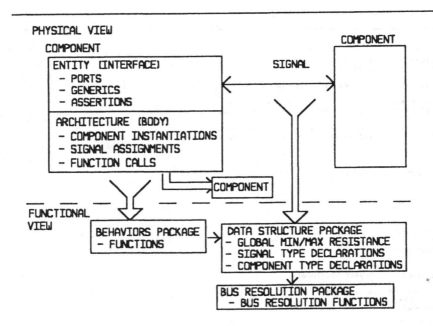

Figure 3.   General VHDL Model Structure

2.2.3   General Model Structure

The model structure of Figure 3 is divided into a physical view and a functional view. The structure applies to transmission line modeling and the analog-digital modeling described in the next chapter.

The two major constructs in the physical view are components and signals. Components represent functional blocks while signals pass information between the functional blocks. Components, entity declarations bound to architectural bodies, define the design structure and reference functions to implement behavior. Signals, the other important physical construct depicted in Figure 3, transfer the behavioral information between components during simulation. They can be of various types including arrays, records, and scalars. As described earlier, multiple signal assignments may drive the same signal (dotting) when a function defining the result is specified.

The functional view of VHDL expressed in packages resembles aspects of a programming language, such as Ada. The general model structure of this chapter (given in Figure 3) uses several packages. The lowest, described earlier, contains bus resolution functions for determining the net result for dotted signals of various types including those defined in the next higher package, the data structure. Again, the data structure package contains signal type definitions, component type definitions for driver signal modes, and the global maximum/minimum resistance values used in transmission line calculations. Finally, the highest level package contains the behavioral functions. The functions use the types defined in the data structure and generate values for signal assignments. These functions and the entities referencing them constitute the VHDL behavioral models.

2.3 BEHAVIORAL MODELS

The basic behavioral models required to simulate transmission line effects are the transmission line (T), driver (D), and receiver (R) shown in Figure 4. These models would replace the digital logic simulation output port, the signal itself, and the input port to signal connection. In general, drivers and receivers buffer the logic output and input ports, respectively, while a network of transmission lines model the signal.

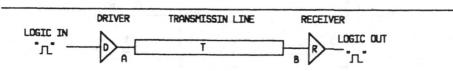

Figure 4. Basic Transmission Line Simulation Models

Each model is implemented as a VHDL entity. At time zero each broadcasts its load conductance to its transmission line node(s). The driver and transmission line models use this loading information to determine the node voltage behavior at A and B. Specifically, the driver model broadcasts its output load conductance to the output node (A) and drives a voltage output after a time delay based on its logic input. The transmission line model broadcasts its load conductance to both its bidirectional nodes (A and B) and voltage drives one side of the line after a time-of-flight (line delay) based on the voltage change at the opposite end of the line. Finally, the receiver model broadcasts its input load conductance to the transmission line node and drives a logic output after a delay based on its voltage input. This completes the logic-in to logic-out data path with transmission line simulation between logic input and output. In this way, combined logic-transmission line simulation may be imple-

mented. Having established the general overview for the models, the
details of the transmission line are next examined.

2.3.1 Lossless Transmission Line

The transmission line model is the most important element of the
total model. It transfers information between nodes bidirectionally,
calculates reflections, and forms the dominant characteristic of the
network. The voltage algorithm is bidirectional, but only one directional path is diagrammed and shown in Figure 5. The reverse direction can easily be derived.

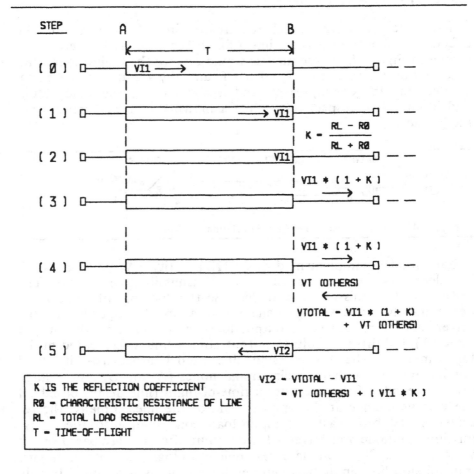

Figure 5.   Transmission Line Algorithm for directional Node A-> B Path

The diagram depicts the dynamics of the directional Node A-> B data path beginning with an incident wave (VI1) at the near-end of the line A traveling toward the far-end of the line B in step 0. This incident wave (VI1) arrives at Node B after a time-of-flight (T) making it internally incident at B in Step 1. This wave (VI1) then encounters a resistance mismatch characterized by the reflection coefficient (K) at the far-end B in step 2. It transmits (1 + K) times its value to all components connected to this line via the external node at B in step 3. This transmitted voltage, superimposed with all the other transmitted voltage waves from the other components connected to the line via the external node at B, forms the total voltage at the far-end B in step 4. Finally, subtracting the incident voltage (VI1) from the total voltage of the external node at B generates the new incident wave traveling toward A (VI2) in step 5 repeating this dynamic cycle for the other direction, the Node B-> A path. The data path structure of this voltage algorithm using the structures defined in the transmission line data structure package is shown in Figure 6.

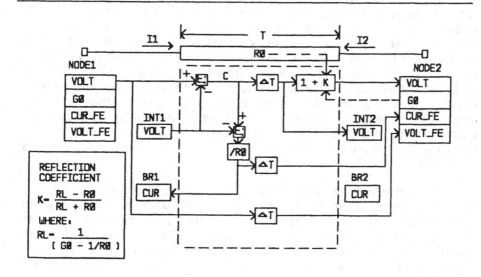

Figure 6.   Transmission line Data Path Flow for Node1-> 2

The dynamic cycle of Figure 6 corresponds directly to the algorithm of Figure 5. It begins with an event at NODE1 which corresponds to step 4 of a Node B-> A cycle. This event occurs at NODE1 representing a new total voltage contained in the NODE1 volt field. Subtracting the incident voltage wave value in the INT1 volt field from this total voltage, generates the new incident wave (C) traveling toward the far-end of the line, NODE2. Wave C begins at the near-end NODE1 and travels to the far-end NODE2 which corresponds to step 0 in Figure 5 for a Node A->B cycle. Wave C propagates to the volt field of INT2 at the far-end of the line after a time-of-flight (T) corresponding to step 1 in Figure 5. Then, comparing the NODE2 load conductance field (G0) to the characteristic resistance generic (R0) of the line determines the reflection coefficient (K). The value of C times $(1 + K)$ is transmitted to all components connected to this line via the external NODE2 volt field which corresponds to step 3 in Figure 5. Finally, in the subsequent delta time processing this transmitted voltage is superimposed with the other voltages driven to the node by the summing resolution function. This generates a new event repeating the cycle of Figure 6 for the Node 2->1 path which corresponds to step 4 in Figure 5. Other quantities are updated based on these transmitted and internally incident voltages.

The total current entering the line at *NODE1* (BR1) is simply the total incident voltage traveling toward *NODE2* (C) less the internally incident voltage at *NODE1* (INT1) divided by the characteristic resistance (R0). This same quantity propagates to the NODE2 cur_fe field after a time-of-flight (T). Similarly, the total voltage (V) propagates to the NODE2 volt_fe field after a time-of-flight (T). The far-end quantities *cur_fe* and *volt_fe* store the total voltage and branch current present at the far-end when the transmitted voltage was determined. These quantities are used when nonlinear elements are connected to the line. The VHDL transmission line model implementing the algorithm and data structure of the previous figures is shown in Figure 7.

The transmission line model entity declaration contains the parameters and interface for the model. The ports are INOUT enabling the bidirectional data flow characteristic of the transmission line. The characteristic resistance (R0) and time-of-flight (T) for the line are specified by customizing the VHDL generics *r_char* and *t_flight*, respectively. Finally, an assertion verifies a positive characteristic resistance for *r_char*.

The architectural body contains the algorithm of the model. At time zero the transmission line conductance, derived from *r_char*, is broadcast to the *g0* field of both transmission line nodes: *node1* and *node2*. Next, it defines two processes, each sensitive to a voltage event on one of the two ports to implement bidirectionality. For a *node1* voltage event, *node2*, *int2*, and *br1* are all updated. Similarly, for a *node2* voltage event, *node1*, *int1*, and *br2* are updated. Signal assignments are made after a time-of-flight, *t_flight*, except for *br2*. Each signal assignment implements one data path from Figure 6. The functions calculating the new signal values execute the transmission line data path representation bounded by the dashed box of Figure 6. The first of these is shown in Figure 8.

```
USE work.tline_data_structure.all;
USE work.tline_active_behaviors.all;

-- interface of transmission line model
ENTITY tline IS
    GENERIC ( r_char  : real    :=  0.080; -- kohms
              t_flight: time    := 10000 ps      );
    PORT    (  node1  : INOUT   tline_node := (volt=>0.0, g0=>g_min,
                                               volt_fe=>0.0, cur_fe=>0.0);
               node2  : INOUT   tline_node := (volt=>0.0, g0=>g_min,
                                               volt_fe=>0.0, cur_fe=>0.0));

BEGIN

    ASSERT( r_char > 0.0 )
        REPORT "ERROR: tline R0 must be > 0.0"
        SEVERITY failure;

END tline;

-- body of transmission line model
ARCHITECTURE tline_behavior OF tline IS
    SIGNAL int1, int2 : tline_traveler := ( volt => 0.0 );
    SIGNAL br1, br2 : tline_branch := ( cur => 0.0 );
    BEGIN
    node1.g0 <= ( 1.0/(r_char + r_min) );
    node2.g0 <= ( 1.0/(r_char + r_min) );

P1:  PROCESS(node1.volt)
     BEGIN
        node2.volt <= TRANSPORT transmit_volt (node1, int1, node2,r_char)
                                               AFTER t_flight;
        node2.volt_fe <= TRANSPORT transmit_vfe (node1)
                                               AFTER t_flight;
        node2.cur_fe  <= TRANSPORT update_current (node1, int1, r_char)
                                               AFTER t_flight;
        int2.volt     <= TRANSPORT propagate (node1, int1)
                                               AFTER t_flight;
        br1.cur       <= TRANSPORT update_current (node1, int1, r_char);

     END PROCESS p1;
P2:  PROCESS(node2.volt)
     BEGIN
        node1.volt <= TRANSPORT transmit_volt (node2, int2, node1,r_char)
                                               AFTER t_flight;
        node1.volt_fe <= TRANSPORT transmit_vfe (node2)
                                               AFTER t_flight;
        node1.cur_fe  <= TRANSPORT update_current (node2, int2, r_char)
                                               AFTER t_flight;
        int1.volt     <= TRANSPORT propagate (node2, int2)
                                               AFTER t_flight;
        br2.cur       <= TRANSPORT update_current (node2, int2, r_char);

     END PROCESS p2;
--
END tline_behavior;
```

Figure 7.   Transmission line Model

```
--
-- propagate function for transmission line segment
--
    FUNCTION propagate      ( node_in :   tline_node   ;
                              int_in  :   tline_traveler
                                        ) RETURN real IS

    BEGIN
        RETURN (node_in.volt - int_in.volt);
    END propagate;
```

Figure 8. Transmission line Propagate Function

For the Node 1->2 direction of Figure 6 the *propagate* function returns the far-end internally incident voltage (C) by taking the difference between the total voltage (NODE1.VOLT) and the internally incident voltage (INT1.VOLT) at the input. This difference constitutes the voltage internally reflected and entering the line at the input node (NODE1). This quantity is also used in the *transmit_volt* function shown in Figure 9.

```
--
-- transmit voltage function for transmission line segment
--
    FUNCTION transmit_volt( node_in :   tline_node   ;
                            int_in  :   tline_traveler;
                            node_out:   tline_node   ;
                            res     :   real         ) RETURN real IS

        VARIABLE drive: real    ; -- returned
        VARIABLE gload: real    ; -- load conductance on driver output
        VARIABLE rload: real    ; -- load resistance  on driver output
        VARIABLE gamma: real    ; -- reflection coefficient

    BEGIN
        gload := node_out.g0 - ( 1.0/(res + r_min) );
        rload := 1.0/(gload + g_min);
        gamma := (rload - res)/(rload + res);
        RETURN  (node_in.volt - int_in.volt) * (1.0 + gamma);
    END transmit_volt;
```

Figure 9. Transmit Voltage Function

In the *transmit_volt* function, the far-end incident quantity (C) is multiplied by the transmission coefficient *(1 + gamma)* yielding the transmitted voltage (NODE2.VOLT).

```
--
-- branch current update function for transmission line segment
--
  FUNCTION update_current(node_in :   tline_node;
                          int_in  :   tline_traveler;
                          res     :   real          ) RETURN real IS

    BEGIN
        RETURN  ((node_in.volt - int_in.volt) - (int_in.volt))
                                /res;
    END update_current;
```

Figure 10. Transmission line Update Current Function

Finally, for the Node 1->2 path the **update_current** function of Figure 10 returns the branch current (BR1.CUR) at the "event" node (NODE1).

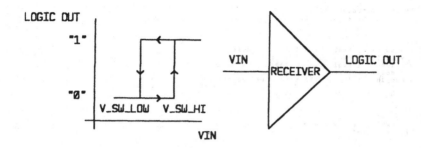

Figure 11.   Receiver Transfer Curve and Block Diagram

2.3.2 Receiver

The hysteresis receiver depicted in Figure 11 exhibits several static characteristics. First, the receiver has two switching levels. The higher one, V_SW_HI, is active when the output is low and the lower one, V_SW_LOW, is active when the the output is high. With V_SW_HI > V_SW_LOW the voltage difference between these two levels constitutes the amount of hyteresis. The receiver also presents an input load conductance to the transmission line and has time delay from input to output.

Dynamically, the receiver exhibits hysteresis switching characteristics. The output switches to a logic "1" state after a delay when the output is in a logic "0" state and the input voltage exceeds the higher threshold. Conversely, the output switches to a logic "0" state after a delay when the output is in a logic "1" state and the input voltage falls below the lower threshold. Finally, the output remains in the same state if neither condition is satisfied.

The VHDL model for the receiver incorporating hysteresis is shown in Figure 12.

```
USE work.tline_data_structure.all;
USE work.tline_active_behaviors.all;

-- interface
ENTITY receiver IS
    GENERIC ( v_sw_hi  : real    := 1.0  ; -- volts
              v_sw_low : real    := 1.0  ; -- volts
              r_input  : real    := 0.080; -- kohms
              t_delay  : time    := 1 ns         );
    PORT    ( rec_in   : INOUT  tline_node := (volt=>0.0, g0=>g_min,
                                              volt_fe=>0.0, cur_fe=>0.0);
              rec_out: BUFFER bit          );
  BEGIN

        ASSERT( r_input > 0.0 )
          REPORT "ERROR: receiver Rin must be > 0.0"
          SEVERITY failure;

END receiver;

-- body
ARCHITECTURE receiver_behavior OF receiver IS
    BEGIN
      rec_in.g0  <= ( 1.0 / (r_input + r_min) );

      rec_out <= '1' AFTER t_delay WHEN (rec_in.volt > v_sw_hi ) ELSE
                 '0' AFTER t_delay WHEN (rec_in.volt < v_sw_low) ELSE
                 rec_out AFTER t_delay;

--
END receiver_behavior;
```

Figure 12.   Receiver Model

The receiver entity declaration defines the receiver model. The output port, *rec_out*, is a simple logic type; however, the input port, *rec_in*, is INOUT enabling the driving of the input load conductance as well as the reading of the input voltage. Several VHDL generics customize each receiver. The switch levels are set by *v_sw_hi* and *v_sw_low*. The input resistance and receiver delay are customized through *r_input* and *t_delay*, respectively. Finally, an assertion verifies a positive value for *r_input*.

In the receiver architectural body, the model first broadcasts its conductance based on *r_input* to its transmission line node. Then, with an event on the transmission line node, it compares the transmission line node voltage to the generic switching levels, *v_sw_hi* and *v_sw_low*, specified by the user. When the transmission line voltage exceeds the high switching threshold, a logic "1" output event for the receiver is scheduled after the user specified delay, *t_delay*. Similarly, when the voltage falls below the low switch level, a logic "0" event is scheduled. Finally, for a level between the two switching levels, the previous value is retained. Having described the transmission line and receiver models, the details of the third fundamental model, the linear driver, are given.

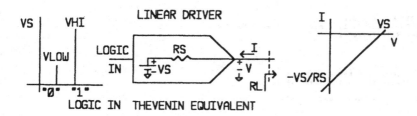

Figure 13.   Linear Driver Transfer Curve, Block Diagram, and I-V Curve

2.3.3 Linear Driver

The linear driver, the structure which is depicted in Figure 13, exhibits several static characteristics. First, the driver has two signal modes, voltage and current. The driver may either drive a high or low value for each mode. Drivers specified in current mode (Norton sources) are converted to equivalent voltage drivers (Thevenin sources) so a voltage is always driven to an external output node. The driver presents a constant output load resistance (RS) to the transmission line and exhibits delay from input to output.

Dynamically, the driver behaves in this manner. The output drives a high voltage level (VS = VHI) after a delay when the input switches to a logic "1". Conversely, the output is driven to a low level (VS = VLOW) when the input switches to a logic "0" state. Regardless of the logic input, the linear driver always presents the same output resistance (RS) to the transmission line.

The VHDL model for the linear driver is shown in Figure 14.

```
USE work.tline_data_structure.all;
USE work.tline_active_behaviors.all;

-- interface
ENTITY linear_drv IS
    GENERIC ( hi      : real      := 5.0  ; -- volts
              low     : real      := 0.0  ; -- volts
              d_kind  : sig_mode  := v    ;
              r_out   : real      := 0.080; -- kohms
              t_delay : time      := 1000 ps     );
    PORT    ( drv_in  : IN    bit;
              drv_out : INOUT  tline_node := (volt  =>0.0, g0    =>g_min,
                                              volt_fe=>0.0, cur_fe=>0.0));

BEGIN

       ASSERT( r_out > 0.0 )
          REPORT "ERROR: linear_drv Rout must be > 0.0"
          SEVERITY failure;

END linear_drv;

-- body
ARCHITECTURE linear_drv_behavior OF linear_drv IS
    SIGNAL listen_toggle : boolean := false;

    SIGNAL drv_in_delay  : bit := '0';

    BEGIN
    drv_out.g0 <= ( 1.0/(r_out + r_min) );
    drv_in_delay <= drv_in AFTER t_delay;

P1:  PROCESS(drv_out.volt, drv_in_delay, listen_toggle)

       VARIABLE this_driver   : boolean := false;

       BEGIN
       IF ( NOT this_driver ) THEN
           drv_out.volt <= drv_linear(drv_in_delay, r_out, drv_out,
                                      low, hi, d_kind);
           this_driver := TRUE;
           listen_toggle <= NOT listen_toggle;
       ELSE
           this_driver := FALSE;
       END IF;
    END PROCESS p1;
--
END linear_drv_behavior;
```

Figure 14.   Linear Driver Model

The linear driver model entity declaration contains the interface. The input port, **drv_in,** is a simple logic type; however, the driver output port, **drv_out,** is INOUT enabling the reading of the transmission line load conductance as well as the driving of an output voltage and driver output conductance (1/RS). Several VHDL generics customize each driver. The linear driver with output resistance drives two levels of either a voltage or current. For the driver of Figure 13, the voltage/current is determined by **d_kind** and selecting between the two user specified levels, **hi** or **low,** is a function of the logic input. The output resistance and delay are customized by **r_out** and **t_delay,** respectively. Finally, an assertion verifies a positive values of **r_out.**

The architectural body implements the behavior. This driver broadcasts its conductance (reciprocal of *r_out*) at time zero to the G0 field of its output node. It applies a voltage to the output each time an event occurs at its input after a time delay, *t_delay*. It also applies a voltage to the output each time an element *other than itself* causes a voltage event at the output node. The driver does not apply a voltage to the output in reaction to an event it caused or an oscillatory condition would result. To filter the driver from its own output node updates, the if-then-else structure in process *P1* is used. This implementation desensitizes the driver for the delta-time that it drives the output node. *Since the semantics for delta-time processing are defined in VHDL, this structure should be simulator independent.* With the correct driver reaction to input stimulus established, the details of the driver function (i.e. determining the output voltage) is next examined.

```
--
--
-- function for linear transmission line driver
--
   FUNCTION drv_linear ( din :   bit;
                         res :   real ;
                         dout:   tline_node;
                         low :   real     ;
                         hi  :   real     ;
                         drv :   sig_mode   ) RETURN real IS

       VARIABLE gload: real      ; -- load conductance on driver output
       VARIABLE rload: real      ; -- load resistance  on driver output
       VARIABLE d_val: real      ; -- driver value
   BEGIN
        gload    := dout.g0 - (  1.0/(res + r_min) );
        rload    := 1.0/(gload + g_min);
        IF (din = '0') THEN
          d_val := low;
        ELSE
          d_val := hi;
        END IF;
        IF (drv = v) THEN
          RETURN  d_val * ( rload/(rload + res) );
        ELSE
          RETURN  d_val * ( 1.0/dout.g0 );
        END IF;
   END drv_linear;
```

Figure 15.   Linear Driver Function

For the driver depicted in Figure 13, the *drv_linear* function in Figure 15 first determines the load resistance from the G0 field of the driver output node. It selects the *hi* value of voltage (VS) or current for a logic '1' input and the *low* value for a logic '0' input. When the value specified is a voltage, voltage division between the driver (RS) and the load (RL) determines the returned voltage. Conversely, when the value specified is a current, it is multiplied by the total node resistance and returned as a voltage.

The linear driver is adequate for Thevenin equivalent drivers and resistive terminations. Basic networks may be simulated by combining this driver with the transmission line and receiver behaviors. However, extensions to this driver model are required for "real" drivers. They exhibit nonlinear behavior and varying resistance requiring a more general driver algorithm.

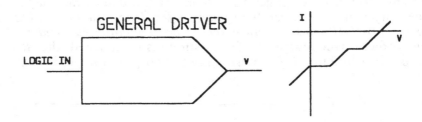

Figure 16.   General Driver Block Diagram and general I-V Curve

2.3.4 General Driver

The theory for the general driver depicted in Figure 16 is based on Bergeron's method [3]. This more general algorithm does not use the driver reflection coefficient; rather it uses the total I-V characteristics of the driver and transmission line. Hence, this general driver model applies to both linear and nonlinear I-V characteristic drivers.

The algorithm consists of several steps. The driver model defines one I-V characteristic, but must obtain the transmission line characteristic resistance along with the far-end voltage and current to determine the transmission line I-V characteristic. This information is available from the driver model output node. Having determined both I-V characteristics, two equations and two unknowns are established. The method next determines the intersection of the two I-V characteristics for the same current by applying Kirchoff's Current Law. Then the equation is solved for the driver output voltage. Note that this resulting equation describes the driver behavior only with a single transmission line load.

To illustrate this general algorithm, the VHDL model for the linear driver of Figure 14 may be used. The graphical depiction and equations for determining the functional result are shown in Figure 17.

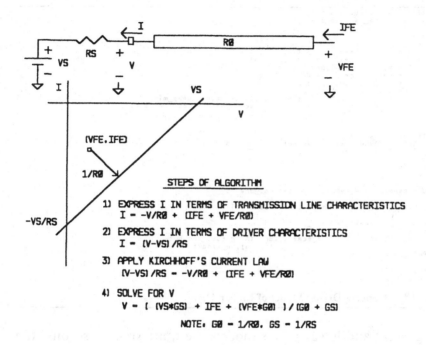

STEPS OF ALGORITHM

1) EXPRESS I IN TERMS OF TRANSMISSION LINE CHARACTERISTICS
   I = -V/R0 + (IFE + VFE/R0)

2) EXPRESS I IN TERMS OF DRIVER CHARACTERISTICS
   I = (V-VS)/RS

3) APPLY KIRCHHOFF'S CURRENT LAW
   (V-VS)/RS = -V/R0 + (IFE + VFE/R0)

4) SOLVE FOR V
   V = ((VS*GS) + IFE + (VFE*G0))/(G0 + GS)
   NOTE: G0 = 1/R0, GS = 1/RS

Figure 17. General Driver Method

The method uses the I-V characteristics of both driver and transmission line to determine the output voltage (V) and current (I). First, the current (I) is expressed in terms of the characteristic resistance (R0), far-end voltage (VFE), far-end current (IFE), and driver voltage (V) in step 1. Next, the method expresses the current (I) in terms of driver source voltage (VS), output resistance (RS), and driver output voltage (V) in step 2. Applying Kirchoff's current law determines the intersection of the two I-V curves at the same current in step 3. This enables solving for the driver output voltage (V) in step 4.

This voltage is then directly applied to the transmission line node connecting the driver output and the transmission line. The general driver model implementation of this linear driver model is shown in figure 18.

```
USE work.tline_data_structure.all;
USE work.tline_active_behaviors.all;

-- interface
ENTITY general_drv IS
    GENERIC ( hi       : real    := 5.0  ; -- volts
              low      : real    := 0.0  ; -- volts
              d_kind   : sig_mode := v  ;
              r_out    : real    := 0.080; -- kohms
              t_delay  : time    := 1 ns);
    PORT    ( drv_in   : IN      bit;
              drv_out  : INOUT   tline_node := (volt=>0.0, g0=>g_min,
                                                volt_fe=>0.0, cur_fe=>0.0));

BEGIN

    ASSERT( r_out > 0.0 )
        REPORT "ERROR: tline_drv Rout must be > 0.0'"
        SEVERITY failure;

END general_drv;

-- body
ARCHITECTURE general_drv_behavior OF general_drv IS
    BEGIN
        drv_out.g0 <= g_max; -- short it so tline drives 0.0v

P1: PROCESS(drv_out.volt_fe, drv_in)
        BEGIN
            drv_out.volt <= drive_general(drv_in, r_out, drv_out,
                                          low, hi, d_kind);
        END PROCESS p1;
--
END general_drv_behavior;
```

Figure 18.   General Driver Model of Linear Driver

The general and linear driver models are quite similar, so only the differences are described. First, the driver model is sensitive to far end voltage changes, so the Boolean signals and the if-then-else construct are no longer required to filter the driver from its own updates. Next, the general driver architecture broadcasts the conductance, **g_max**, at time zero to the transmission line node. This presents a "short" to the transmission line yielding a near-zero transmission coefficient for voltage. Hence, effectively zero volts is applied to the node by the transmission line. This zero voltage summed with the result of the driver function through a summing resolution function yields the total node voltage. Thus, to obtain the correct result the **total** voltage must be returned by the driver function of Figure 19 for application to the driver output node.

The general and linear driver functions are also similar in many respects, but two key differences exist. First, the *drive_general* function calculates the load resistance from the G0 field of the driver output by subtracting GMAX. Second, this function uses the relation for driver output voltage derived in Figure 17 to determine the correct total voltage output value. This is in contrast to the voltage division used in the linear driver function of Figure 15.

```
--
  --
  -- function for transmission line driver using general voltage
  --
      FUNCTION drive_general ( din : bit;
                               res :  real ;
                               dout:  tline_node;
                               low :  real     ;
                               hi  :  real     ;
                               drv :  sig_mode    ) RETURN real IS

         VARIABLE gload: real      ; -- load conductance on driver output
         VARIABLE gdrv : real      ; -- load resistance  on driver output
         VARIABLE d_val: real      ; -- driver value
         VARIABLE v_int: real      ; -- voltage source intercept

      BEGIN
          gdrv  := 1.0/(res + r_min);
          gload := dout.g0 - g_max + g_min;
          IF (din = '0') THEN
             d_val := low;
          ELSE
             d_val := hi;
          END IF;
          IF (drv = v) THEN
             v_int := d_val;
          ELSE
             v_int := d_val * res ;
          END IF;
          RETURN    ( dout.cur_fe + (dout.volt_fe * gload) +
                      (v_int * gdrv) )/(gload + gdrv);
      END drive_general;
  --
```

Figure 19.   Drive_general Function for General Driver Model

2.4 APPLICATION OF TRANSMISSION LINE BEHAVIORS

Transmission line networks have been constructed by combining the previously defined models. Specifically, using the general driver model, FET push-pull, open-drain, and other nonlinear drivers have been modeled. Even iteration is easily accomplished in the function to determine effects such as body effect in FET n-channel output stages.

2.4.1 FET Modeling with the General Driver Model

The FET equations used for an n-channel device are the standard first order I-V equations for the linear and saturated modes of operation [5]. Unlike the work described in the previous chapter on "Switch Level Modeling in VHDL", these models incorporate the full floating point accuracy of the model rather than abstract it to the digital logic levels (0, 1, X, etc.,). This enables modeling the nondiscrete aggregate effect of incident, reflected, and transmitted voltages and currents which determine transmission line behavior. Also, this method does not focus on pass transistors; however, it does apply local relaxation.

The general driver method applied to an n-channel FET open-drain driver is identical to the one depicted in Figure 17 for the linear driver. The driver equations differ and two regions of operation must be modeled as shown in Figure 20.

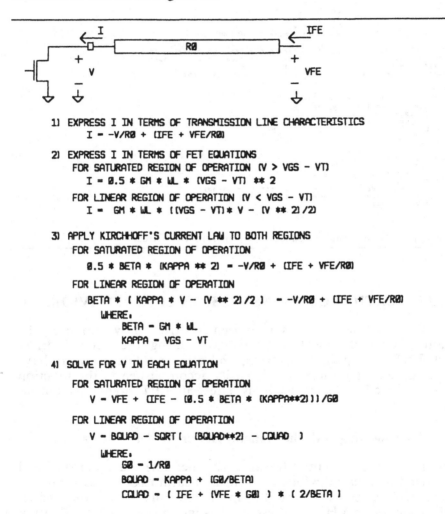

1) EXPRESS I IN TERMS OF TRANSMISSION LINE CHARACTERISTICS
   I = -V/R0 + (IFE + VFE/R0)

2) EXPRESS I IN TERMS OF FET EQUATIONS
   FOR SATURATED REGION OF OPERATION (V > VGS - VT)
       I = 0.5 * GM * WL * (VGS - VT) ** 2

   FOR LINEAR REGION OF OPERATION (V < VGS - VT)
       I = GM * WL * ((VGS - VT) * V - (V ** 2) /2)

3) APPLY KIRCHHOFF'S CURRENT LAW TO BOTH REGIONS
   FOR SATURATED REGION OF OPERATION
       0.5 * BETA * (KAPPA ** 2) = -V/R0 + (IFE + VFE/R0)

   FOR LINEAR REGION OF OPERATION
       BETA * (KAPPA * V - (V ** 2) /2) = -V/R0 + (IFE + VFE/R0)
           WHERE:
                 BETA = GM * WL
                 KAPPA = VGS - VT

4) SOLVE FOR V IN EACH EQUATION
   FOR SATURATED REGION OF OPERATION
       V = VFE + (IFE - (0.5 * BETA * (KAPPA**2)))/G0

   FOR LINEAR REGION OF OPERATION
       V = BQUAD - SQRT ((BQUAD**2) - CQUAD)
           WHERE:
                 G0 = 1/R0
                 BQUAD = KAPPA + (G0/BETA)
                 CQUAD = (IFE + (VFE * G0)) * (2/BETA)

Figure 20.   General Driver Method for FET N-channel Open-drain Driver

The n-channel FET open-drain model uses the I-V characteristics of both driver and transmission line to determine the output voltage (V) and current (I). Again, the current (I) is expressed in terms of the characteristic resistance (R0), far-end voltage (VFE), far-end current (IFE), and driver voltage (V) in step 1. Next, the method expresses the current (I) in terms of the FET transconductance (GM), width-to-length aspect ratio (WL), base threshold voltage (VT), FET input voltage (VGS), and FET output voltage (V) for each region of operation in step 2. Applying Kirchoff's current law to both regions determines the intersection of the two I-V curves at the same current in step 3 which enables solving for the driver output voltage (V) in step 4 of both regions for a single transmission line load. The VHDL model for the n-channel open-drain driver is shown in Figure 21.

```
USE work.tline_data_structure.all;
USE work.tline_active_behaviors.all;

-- interface
ENTITY fet_od_drv IS
      GENERIC ( gm        : real    := 5.0   ; -- a/(v*v)
                wl        : real    := 0.0   ;
                vt        : real    := 0.0   ; -- volts
                vgs       : real    := 0.0   ; -- volts
                t_delay : time    := 100 ns);
      PORT    ( din       : IN  bit;
                dout      : INOUT  tline_node := (volt=>0.0, g0=>g_min,
                                                 volt_fe=>0.0, cur_fe=>0.0));

BEGIN

      ASSERT( (gm > 0.0) AND (wl > 0.0) AND (vt > 0.0) AND (vgs > 0.0)  )
         REPORT "ERROR: tline_fet_drv generics must be > 0.0"
         SEVERITY failure;

END fet_od_drv;

-- body
ARCHITECTURE fet_od_drv_behavior OF fet_od_drv IS
      BEGIN
         dout.g0 <= g_max; --short it so tline drives 0 volts

P1:    PROCESS(dout.volt_fe, din)
         BEGIN
             dout.volt <= drive_fet(din, dout, gm, wl, vt, vgs);

         END PROCESS p1;
--
END fet_od_drv_behavior;
```

Figure 21.   General Driver Model of FET N-Channel Open-drain Driver

The generalized FET and linear driver models are similar. Aside from the specific generics used by each driver, the models are the same. Both the FET and linear driver architectures broadcast conductance **g_max** at time zero to the transmission line node which presents a "short" to the transmission line yielding a near-zero transmission coefficient for voltage. Hence, effectively zero volts is applied to the node by the transmission line. The zero voltage summed with the result of the driver function, **drive_fet**, through a summing resolution function yields the total node voltage. Thus, to obtain the correct result the **total** voltage must be returned by the FET function shown in Figure 22.

```
--
--
-- function for open drain transmission line FET driver
--
    FUNCTION drive_fet      ( din :    bit       ;
                              dout:    tline_node;
                              gm  :    real      ;
                              wl  :    real      ;
                              vt  :    real      ;
                              vgs :    real      ) RETURN real IS

        VARIABLE gload: real; -- load conductance on driver output
        VARIABLE d_val: real; -- driver value
        VARIABLE kappa: real; -- voltage overdrive on fet
        VARIABLE beta : real; -- transconductance of fet
        VARIABLE bquad: real; -- b term in quadratic equation
        VARIABLE cquad: real; -- c term in quadratic equation
        VARIABLE disc : real; -- b**2 - c
        VARIABLE sat  : real; -- saturation current of fet

    BEGIN
        gload := dout.g0 - g_max + g_min;
        IF (din = '0') THEN
         d_val := dout.volt_fe + (dout.cur_fe/gload);--hiz
        ELSE
         kappa :=   vgs - vt;
         beta  :=   gm  * wl;
         sat   := (beta/2.0) * (kappa * kappa);
         d_val := dout.volt_fe + (dout.cur_fe - sat)
                                    /gload; -- saturated
         IF (d_val < kappa) THEN
           bquad := kappa + (gload/beta);
           cquad := ( dout.cur_fe + (dout.volt_fe * gload) )
                              *(2.0/beta);
           disc  := (bquad * bquad) - cquad;
           d_val := bquad - sqrt(disc); -- linear
         END IF;
        END IF;
        RETURN  d_val;
    END drive_fet;
--
```

Figure 22.   Drive_fet Function for General Driver Model

The FET and Linear driver functions are also similar in many respects, but one key difference exists. The FET function determines the correct region of operation -- linear, saturated or "OFF". The "OFF" or "hiz" region is enabled by a logic "0" on the Boolean input, **din**. For a logic "1" the linear or saturated operation is determined by consistency between assumption and resulting value. The function assumes the saturated mode of operation when enabled and then checks the resulting voltage. If the resulting voltage is in the linear region, its value is recalculated with the quadratic linear region equation before it is returned. To solve the quadratic, the square root function *sqrt* shown in Figure 23 is used.

```
--
--funciton for square root
--
  FUNCTION sqrt( x: real ) RETURN real IS

    VARIABLE  sqrtx : real;

  BEGIN
        IF x = 0.0 THEN
          RETURN 0.0;
        ELSE
          sqrtx := x;
          FOR i IN 1 TO 20 LOOP
            sqrtx := (sqrtx + (x/sqrtx)) * 0.5;
          END LOOP;
        END IF;
        RETURN sqrtx;
  END sqrt;
--
```

Figure 23.   Square Root Function

This simple function approximates the square root of *x* through successive iteration.

2.4.2 Network Example

These models combined form the basis for simulating memory and logic card designs where more complicated transmission line characteristics are needed. An example network using these models is shown in Figure 24.

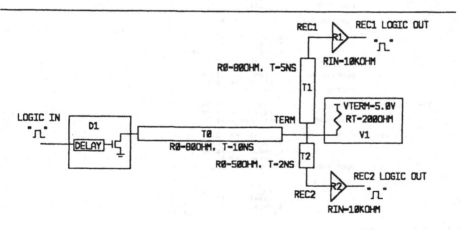

Figure 24.   FET Driver Example

The transmission lines (T0-T2), receivers (R1-R2), and both driver type models (D1 and V1) are used. The terminator (V1) is modeled as a linear driver and the open-drain FET driver (D1) is modeled as a general driver using its high impedance "OFF" state and fundamental two-region current equation. The VHDL model for this simulation example is shown in Figure 25. For brevity, the binding indications are omitted from the declarative region.

```
USE work.tline_data_structure.all;

--interface
ENTITY main_network IS
END main_network;

--body
ARCHITECTURE main_network_sample OF main_network IS
        SIGNAL drv_logic_in   : bit := '0';
        SIGNAL rec1_logic_out : bit := '0';
        SIGNAL rec2_logic_out : bit := '0';
        SIGNAL term_in        : bit := '0';
        SIGNAL drv_out : tline_node := ( volt    => 0.0, g0      => g_min,
                                         volt_fe => 0.0, cur_fe  => 0.0   );
        SIGNAL rec_1   : tline_node := ( volt    => 0.0, g0      => g_min,
                                         volt_fe => 0.0, cur_fe  => 0.0   );
        SIGNAL rec_2   : tline_node := ( volt    => 0.0, g0      => g_min,
                                         volt_fe => 0.0, cur_fe  => 0.0   );
        SIGNAL term    : tline_node := ( volt    => 0.0, g0      => g_min,
                                         volt_fe => 0.0, cur_fe  => 0.0   );
-- Partial Declarative Region
  --instantiate
    BEGIN

        term_in       <= '1' AFTER 100 ps;
        drv_logic_in  <= '1' AFTER 200 ns, '0' AFTER 400 ns;

        D1: fet_od_drv   GENERIC MAP (  gm       =>    0.02,
                                        wl       => 200.0 ,
                                        vt       =>    1.0 ,
                                        vgs      =>    5.0 ,
                                        t_delay  => 1 ns    )
                         PORT MAP    (  drv_logic_in, drv_out );

        T0: tline        GENERIC MAP (  r_char   =>   0.08,
                                        t_flight => 10 ns   )
                         PORT MAP    (  drv_out, term );

        V1: linear_drv   GENERIC MAP (  hi       =>   5.0  ,
                                        low      =>   0.0  ,
                                        d_kind   =>   v    ,
                                        r_out    =>   0.200,
                                        t_delay  =>   100 ps )
                         PORT MAP    (  term_in, term);

        T1: tline        GENERIC MAP (  r_char   =>   0.08,
                                        t_flight =>  5 ns   )
                         PORT MAP    (  term, rec_1 );

        T2: tline        GENERIC MAP (  r_char   =>   0.05,
                                        t_flight =>  2 ns   )
                         PORT MAP    (  term, rec_2 );

        R1: receiver     GENERIC MAP (  v_sw_hi  =>   2.5,
                                        v_sw_low =>   2.5,
                                        r_input  => 10.0,
                                        t_delay  => 1 ns    )
                         PORT MAP    (  rec_1, rec1_logic_out );

        R2: receiver     GENERIC MAP (  v_sw_hi  =>   2.5,
                                        v_sw_low =>   2.5,
                                        r_input  => 10.0,
                                        t_delay  => 1 ns    )
                         PORT MAP    (  rec_2, rec2_logic_out );
--
END main_network_sample;
```

Figure 25.   FET Network Entity

The component instantiations correspond directly to Figure 24. Transmission lines T0 and T1 are 80 OHM lines with 10 ns and 5 ns time-of-flights, respectively. Similarly, T2 is a 50 OHM line with a 2 ns time-of-flight. Both receivers R1 and R2 are characterized by 10 Kohm input resistance and a single 2.5 volt switching threshold. Hysteresis is not used to emphasize the network false switching detection. FET driver D1 is characterized by a 20 uA/V**2 transconductance (GM), a width to length aspect ratio (WL) of 200, a 1 volt base threshold voltage (VT), a 5 volt input voltage (VGS), and a delay of 1 ns. The linear driver modeling the terminator (V1) applies 5 volts after 100 ps and presents a constant 200 ohms output resistance to the line. This switching activity is controlled by the **term_in** signal assignment. Finally, the FET is switched by the **drv_logic_in** signal assignment turning it "ON" at 200 ns and turning it "OFF" at 400 ns.

2.4.3 Simulation Results

The resulting waveforms from the VHDL simulation are shown in Figure 26 along with the ASTAP circuit simulation waveform *NREC1* for the *REC1* input voltage. Note that these two waveforms (*NREC1* and *REC_1.VOLT*) overlap with reasonable accuracy resulting in an average error of less than 3.0%. Results may vary with differing simulation conditions. More importantly, this simulation reveals a false switching behavior at the receiver output, *REC1_LOGIC_OUT*, which would go undetected in traditional logic behavior simulation with delay equations.

2.4.4 Limitations and Usage

The behaviors presented model lossless transmission lines and their interaction with drivers and receivers based on resistive loading effects. Since the resistive loadings are broadcast at time zero, test case stimulus must only be applied after time zero. In addition, the general driver algorithm is currently limited to a fanout of one transmission line. Thus, a driver using it may be attached to transmission line networks only at nodes where the driver is singly loaded. Many networks conform to this restriction so many may be analyzed. Drivers with multiple transmission line loads may be modeled by inserting a very short section of line between the driver output and the fanout of N lines.

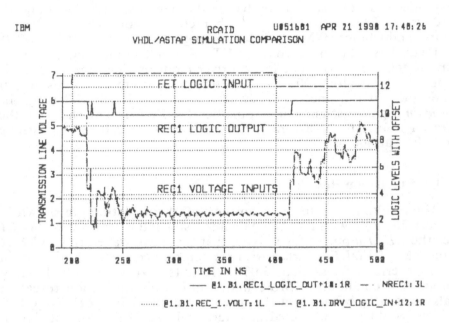

Figure 26. FET Driver Example Waveforms

These models primarily replace logic models for off-chip networks such as those found on cards. The additional aspects modeled by the transmission line behaviors improve the accuracy of the simulation and detect unwanted transmission line aspects.

If parallelism is preferred over replacement, a Boolean field in the transmission line node record may be added. This field may be updated after a time-of-flight by a signal assignment defined in each process of the *tline_behavior*. This would effectively create a Boolean delay line in parallel with the transmission line behavior. Improving simulation accuracy by using these models does increase CPU time for a simulation and may pose problems for synthesis.

The accuracy/runtime trade-off is directly related to the time-of-flight delay used. Specifying line time-of-flights with 100 ps resolution uses the same order of magnitude CPU time as the equivalent SPICE simulation. Runtime increases ten times for 10 ps resolution and decreases 10 times for 1 ns resolution as used in this example. Typically, 100 ps resolution is adequate for digital hardware.

2.5 SUMMARY

In addition to logic behavior, VHDL can model the transmission line behavior for a digital hardware design. Applying the transmission line behaviors to real systems yielded good results. Expanding on the techniques and general models presented, memory/logic card designs have been simulated. These simulations have provided valuable insight early-on in identifying potential problem areas in card layouts. Finally, incorporating this level of detail in a digital simulation model further integrates the simulation environment.

This integration parallels the increased integration in hardware designs and helps satisfy present and future simulation needs. With higher levels of integration, VHDL can model the transmission line effects between chips in multichip modules which are difficult to prototype. Also, as operating frequencies increase, it can model many on-chip transmission line effects nearly impossible to prototype. Finally, with the increased integration of analog-digital circuits, VHDL can also model many of these combined systems as described in the next chapter.

VHDL, rich in the constructs found in both an HDL and a programming language, has the flexibility and structure to model digital designs behaviorally reflecting the transmission line effects associated with interconnection in multichip environments.

REFERENCES

[1] "VHDL User's Manual," No. IR-MD-065-1, Intermetrics Inc., Bethesda, MD, 1985

[2] B. R. Stanisic and M. W. Brown, "VHDL Modeling for Analog-Digital Hardware Designs," in Proc. IEEE International Conference on Computer-Aided Design, November, 1989.

[3] George Metzger, et al., Transmission Lines with Pulse Excitation, Academic Press, New York, NY, 1969.

[4] R. G. Brown, et. al, Lines, Waves and Antennas, John Wiley & Sons, New York, NY, 1973

[5] D. A. Hodges and H. G. Jackson, Analysis and Design of Digital Integrated Circuits, McGraw-Hill, New York, NY, 1983

[6] Advanced Statistical Analysis Program (ASTAP) Program Reference Manual, Pub. No. SH20-1118, IBM Corp., White Plains, NY, 1973.

[7] L. W. Nagel, "SPICE2, A Computer Program to Simulate Semiconductor Devices," ERL Memorandum ERL-M520, University of California, Berkeley, May 1975.

[8] "IEEE Standard VHDL Language Reference Manual", IEEE Standard 1076-1987, New York, NY.

Chapter 3

Behavior Modeling of Mixed Analog-Digital Circuits

Balsha R. Stanisic
and Mark W. Brown

International Business Machines
Rochester, Minnesota

Over the past few years a number of simulation approaches for mixed analog-digital hardware designs have been described. The IEEE standard Hardware Description Language (VHDL), standardized by the IEEE in 1987 and primarily used to model digital behavior, also has the descriptive capability to model analog behavior for event-driven simulation [4]. This chapter describes an approach for mixed analog-digital event-driven simulation and physical design verification using VHDL. Specifically, this chapter presents a general VHDL simulation model and a standard-cell based physical design verification methodology. It demonstrates the applicability of this general simulation model to transfer function modeling through a detailed amplifier example.

This general simulation model and its associated physical design verification methodology were used to verify both the analog-digital functional interaction and the physical interconnection of a Bipolar/CMOS chip (5.5mm x 5.5mm) used in hard disk drive products.

3.1 INTRODUCTION

With higher levels of integration more integrated circuits contain both analog and digital circuitry. Interfacing these circuits on-chip rather than on-card places greater demands on chip simulation since breadboarding the analog-digital interface is not feasible. Hence, a greater emphasis has been placed on mixed analog-digital simulation [4,5,12].

Traditional verification methods applied to hardware interaction and interconnection are inadequate for an analog-digital hardware design. Digital designs are typically modeled behaviorally in a hardware description language (HDL) at the gate or higher level for event-driven simulation. Analog designs are usually modeled structurally in another HDL at the device level for a circuit simulator such as ASTAP (Advanced STatistical Analysis Program) or SPICE. Each design is simulated in its own environment and verified with its own physical design verification methodology. However, these techniques do not use simulation to verify function and important system level quantities such as control-loop stability and feedback timing across the analog-digital interface. Therefore, the integrity of the interconnections at the interface is questionable since combined physical design verification is not done. Only visual inspection and communication between the analog and digital designers verifies the analog-digital interface and is limited to the familiarity of each designer with each HDL.

More recent verification methods applied to hardware interaction improve certification of an analog-digital hardware design. Some methods extend circuit simulation for analog-digital designs through piecewise approximate variable accuracy event-driven simulation [14]. Others have extended the capabilities of a conventional circuit simulation based environment to model digital behavior [10]. Some simulation environments have coupled both digital event-driven and analog circuit simulation [9]. Finally, some event-driven based simulation environments have the capability to model analog function behaviorally [4,5,11].

There are benefits and limitations to each method. Simulation environments restricting the analog partition to conventional circuit simulation exhibit the limitations of circuit simulation and lack the benefits of behavioral simulation. These limitations and benefits are the same as in modeling existing digital designs with many devices. Modeling the design at a gate or higher level is required for total chip simulation since detailed circuit level simulation is CPU intensive and converging on a solution is often difficult [5]. Because full circuit simulation is difficult, such simulators are primarily used to extract timing information for each digital gate or behavior individually. For these same reasons, analog designs comprised of many devices require modeling at a higher level. Here circuit simulators extract the salient features of each analog circuit to create an analog behavior. An efficient mixed analog-digital simulation environment for chips requires an analog behavioral capability.

Recognizing the importance of analog behavioral simulation and the capability of VHDL to describe both analog and digital behavior, the IEEE Standards Coordinating Committee on Analog Hardware Description Languages (SCC 30) has established a working group to examine a VHDL based Analog Hardware Description Language (AHDL) [6]. In addition to proposing the necessary VHDL extensions for an AHDL to support a circuit level descriptive capability, the group is developing VHDL analog behavioral techniques [13].

This chapter describes a general VHDL simulation model for creating a behavioral (not structural) model of an analog circuit and a physical design verification methodology which can alleviate interaction and interconnection verification problems at the analog-digital interface of a hardware design. The behavioral models created are simpler than structural models using primitive components as found in SPICE or ASTAP, however, they can be used to verify the function and system parameters such as control-loop stability and feedback timing. With structural modeling, the corresponding interconnections at the analog-digital interface can be verified. Also, communication between designers is improved by using a common language. These attributes make VHDL an effective modeling language for simulating the analog-digital interaction behaviorally and assuring the correspondence to the analog-digital interconnection.

3.2 SIMULATION MODEL

A simulation model will be described in three parts. The first describes the basic behavioral modeling theory, the second the general model structure, and the third describes the digital and analog applications of the model structure.

3.2.1 Theory

The goal is to behaviorally model lumped-constant electrical networks, typically described in elementary engineering circuit analysis textbooks, within the constraints of the VHDL discrete time environment.

The method used is transfer function modeling. A transfer function is a means of describing the outputs of a component or network as a function of its inputs. This is consistent with the way VHDL models digital behavior because a signal assignment statement essentially implements a digital transfer function. Usually, any component can be modeled if the inputs and outputs are uniquely distinguishable and the external load impedances have only second-order effects on the component behavior. Passive components such as resistors, capacitors and inductors cannot be modeled with this technique because they have no unique inputs and outputs [5].

Generally, a transfer function given in terms of complex frequency, H(S), can be obtained for any linear network through frequency domain analysis using Laplace Transforms. In using frequency domain analysis, transfer functions that contain derivatives with respect to time can be manipulated and simplified by simple algebraic methods. The following single pole filter example shown in Figure 1 illustrates this point.

The transfer function, H(s), in step 1 is obtained by expressing the network elements in terms of the complex frequency, s, and using simple voltage division. This sort of algebraic manipulation also applies to higher order equations. Then step 2 inverse transforms the transfer function resulting in the time domain equation assuming no initial conditions.

As stated in the beginning of this section, these behaviors must be simulated in a discrete time domain. Therefore, the continuous nature of analog signals must be modeled both in magnitude and in time within this discrete environment. The magnitude of these continuously varying signals can be represented with a floating point number in VHDL. The problems associated with time quantization are more difficult. Usually, a piecewise linear approximation provides an accurate representation of a waveform, has no discontinuities and allows use of simple numerical differentiation techniques in evaluating transfer functions.

STEP

1) WRITE TRANSFER FUNCTION H(S)

$$
+ \qquad R \qquad \qquad +
$$
VIN (S)   1/SC   VOUT (S)

$$H(S) = \frac{VOUT\ (S)}{VIN\ (S)} = \frac{1}{1 + SRC}$$

$$VIN\ (S) = VOUT\ (S) + SRC\ VOUT\ (S)$$

2) INVERSE TRANSFORM

$$S \longrightarrow \frac{d}{dt} \quad \text{(NO INITIAL CONDITION)}$$

$$VIN\ (t) = VOUT\ (t) + RC\ \frac{d}{dt}\ VOUT\ (t)$$

3) CONTINUOUS TO DISCRETE

t = NT, T IS GLOBAL TIME INTERVAL

$$\left. \frac{d}{dt}\ VOUT\ (t) \right|_{t=NT} \cong \frac{VOUT\ (NT) - VOUT\ (NT-T)}{T}$$

4) SUBSTITUTE

$$VIN\ (NT) = VOUT\ (NT) + RC \left[\frac{VOUT\ (NT) - VOUT\ (NT-T)}{T} \right]$$

5) SOLVE

$$VOUT\ (NT) = \frac{T\ VIN\ (NT) + RC\ VOUT\ (NT-T)}{T + RC}$$

Figure 1.   Single Pole Filter Transfer Function Example

The numerical differentiation technique used, **Euler's method** [16], is actually the truncated Taylor series shown in Equation (1).

$$\dot{f}(t) = f(t)\Big|_a + \frac{d\ f(t)}{dt}\Big|_a \ (t-a) + \ldots \qquad (1)$$

$$\text{let } T = t-a$$

$$\therefore \ a = t-T$$

Where **T** is the time step between approximations that effectively provides the time quantization. Choosing the value of **T** is very important. It will be discussed later along with the resulting truncation error.

Solving for the derivative and replacing **t** with **NT** provides the approximation shown in equation (2).

```
 d f(t) |           . f(NT) - f(NT - T)
--------|         = ------------------   (2)
   dt   |t=NT               T
```

Using this approximation for the transfer function example results in the continuous to discrete approximation shown in step 3 of Figure 1. Substituting this approximation into the transfer function results in the equation shown in step 4.

Finally, solving for Vout results in the final incremental equation shown in step 5. This provides an incremental equation for determining **Vout** based only on **Vin** and the previous value of **Vout**. This equation, which is encapsulated within a VHDL function subprogram, is used by an amplifier example described later.

Since the Euler method represents a truncated Taylor series, the truncation error is on the order of the next term in the series. For the Euler method, the error is proportional to $T^{**}2$. Therefore, varying T will significantly affect the truncation error [16]. Higher order methods can be used in place of Euler's method, however, their computation cost is higher because of the additional complexity. In a practical sense, for the single pole function above, stability is assured if $T < 2RC$ where RC is the single pole time constant. Further, the truncation error is manageable if T is set in accordance with the smallest time constant within the transfer function being modeled [5].

3.2.2 General Model Structure

The model structure of Figure 2 is similar to the general model structure described in the previous chapter on transmission line modeling. The structure applies to both digital and analog partitions of a hardware design. The physical view is identical to the last chapter. The structure of using multiple packages in the functional view is also identical to the previous chapter; however, the contents of the data structure and behavioral functions (termed behaviors) package differ.

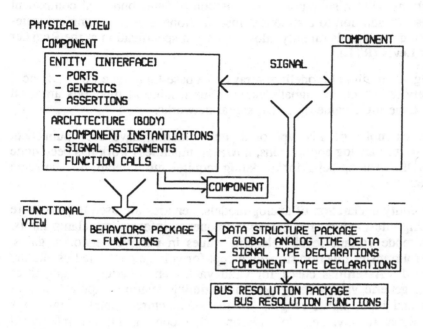

Figure 2. General VHDL Model Structure

The general model of Figure 2 uses several packages. The lowest contains bus resolution functions for determining the net result for dotted signals of various types including those defined in the next higher package, the data structure. The data structure package contains signal type definitions, component type definitions, and a global analog time delta constant used in finite-difference calculations. Finally, the highest level package contains the behavioral functions. These functions use the types defined in the data structure package and generate values for signal assignments.

With the constructs from the physical view and the package structure expressing the functional view, this general model structure can be applied to digital and analog circuits.

3.2.3 Application to Circuits

In digital modeling, certain constructs of VHDL are used in standard ways. For example, in the entity declaration, generics pass timing information such

as rising and falling delay. Assertions check for violations in timing requirements such as setup and hold times. In the architecture, component
instantiations build a structural representation of interconnected component
functions. In addition to component instantiations, signal assignment statements can drive signals directly allowing design specification at the Register
Transfer Level (RTL).

For signals in digital modeling, arrays of a base logic type ('0','1','X',etc.,)
adequately represent the signal's value. Bus resolution functions are used
to resolve the logic value of dotted signal networks.

Many examples of this type of digital modeling have been published.
With respect to analog applications, however, much less work has been done
with VHDL. Analog behavior modeling requires more detail, but the same
techniques apply.

In the entity declaration of analog models, generics are used to pass more
than timing information. For example, frequency information, clamp limits,
common-mode voltage levels, element values in passive networks, gains,
and control transfer function slope/intercept for voltage controlled oscillators
are defined. Assertions check for valid values on a variety of quantities
including generic values being passed-in, timing relations, applied control
voltages and input signal ranges. In the architecture, signal assignments
force a signal to a value. In addition, other components are referenced
through component instantiations. For example, if an amplifier transfer
function requires a pole or zero, a passive network is referenced and customized by passing element values through the entity declaration via a generic parameter. Finally, functions are also called within the architecture.
They are called by a passive component for finite-difference equation calculation and referenced by the top level component to resolve control
transfer curves.

VHDL signals used in the analog partition can be grouped into two categories: *reference* and *varying*. *Reference* signal values do not change with
simulation time. Since there are few reference signals in an analog partition
they are defined with a unique type so any error in their interconnection is
found early in the compilation process. *Varying* signal values change with
simulation time and make up most of the signals in an analog partition. The
signals are defined as record types with a minimum of two fields. The first
field is of type *real* which contains the value of the signal. The second is
an enumeration type indicating the kind of value (eg. voltage, current).

Circuits using *current* type signals often require summing so the resolution feature of VHDL is used. The bus resolution function for these *current* networks sums the value fields of the signal assignments that drive the dotted signal.

An example using the amplifier circuit shown in Figure 3 illustrates some of the techniques outlined above.

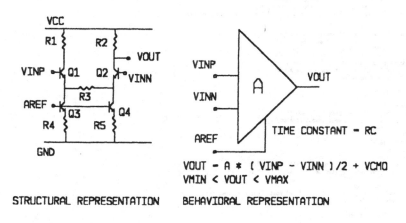

VOUT = A * (VINP - VINN)/2 + VCMO
VMIN < VOUT < VMAX

STRUCTURAL REPRESENTATION   BEHAVIORAL REPRESENTATION

Figure 3.   Basic Amplifier Circuit

The amplifier structural representation is used for physical design verification of the VHDL model behavioral representation. For simulation, the output voltage (VOUT) is the sum of the common-mode output voltage (VCMO) and the gain (A) times half the input voltage difference (VINP - VINN). The output voltage is bounded between the high (VMAX) and low (VMIN) voltage saturation levels and at high frequencies is affected by a single dominant pole (1/RC). The VHDL behavioral model of the amplifier is shown in Figure 4.

The entity declaration for this model defines the ports, generics, and assertions. This component has two *varying* input ports, **vinp** and **vinn**, and one buffer port **vout** of type **analog_signal**. It also has input port *aref*, an analog constant voltage reference [1].

---

[1]   Refer to the associated analog data structure found in Figure 6.

```
USE work.analog_data_structure.all;
USE work.analog_passive_behaviors.all;

-- interface
ENTITY amplifier IS
    GENERIC ( gain     : analog_vv_gain;
              vcmo     : analog_v_level;
              vout_min : analog_v_level;
              vout_max : analog_v_level;
              r        : analog_component;
              c        : analog_component);
    PORT    ( vinp  : IN     analog_signal;
              vinn  : IN     analog_signal;
              aref  : IN     analog_reference_aref;
              vout  : BUFFER analog_signal := (val => 0.0, kind => v));

BEGIN
    ASSERT( (r.val > 0.0) AND (c.val > 0.0) )
        REPORT "ERROR: R and C must be > 0.0"
        SEVERITY failure;

    ASSERT( vout_max.val > vout_min.val )
        REPORT "ERROR: vout_max must be > vout_min"
        SEVERITY failure;

END amplifier;

-- body
ARCHITECTURE amplifier_behavior OF amplifier IS
    SIGNAL ac_in, ideal_vout, pvout: analog_signal :=
           (val => 0.0, kind => v );

-- instansiate
    BEGIN
        -- Ideal Amplifier
        ac_in.val      <= (vinp.val - vinn.val) * 0.5;
        ideal_vout.val <= vcmo.val + ( ac_in.val * gain.val ) ;

        -- Single Pole
        vout       <= rc( r, c, pvout, ideal_vout, vout_min, vout_max);
        pvout      <= vout      AFTER analog_time_delta; -- analog tstep

END amplifier_behavior;
```

Figure 4. Basic Amplifier Model

For the amplifier circuit shown in Figure 3, the generics customize the RC dominant pole roll-off point, the output common-mode voltage level (VCMO), the output saturation limits (VMIN and VMAX), and the gain (A) for the component. The assertion statements check that the component was passed positive values for the resistance (R) and capacitance (C), and that the maximum saturation limit for *vout* is greater than its minimum saturation limit. If an error occurs, the assertion should cause a report to the user and stop the simulation.

The architecture for this component declares and initializes several internal signals. Then, using dataflow (concurrent) equations, it calculates half the applied input voltage difference and assigns it to the value field of signal *ac_in*. When multiplied by the gain and summed with the common-mode voltage of the output, it provides the ideal output voltage delta, *ideal_vout*. Voltage *ideal_vout*, *pvout*, and the component's generics are passed to the function *rc* which assigns the resulting output voltage to *vout*. The next signal assignment takes the present value of *vout* and assigns it to *pvout* so the previous value is stored for the next call. Note also that the assignment to *pvout* in this case is after an *analog_time_delta*. This is the time delta at which analog signals are updated. Finally, the contents of the *rc* function shown in Figure 5 reveals the finite-difference calculation used to generate the next value for *vout*.

```
..
..
-- function for RC - Low Pass
..
FUNCTION rc ( r             : analog_component;
              c             : analog_component;
              pred1_vout    : analog_signal;
              vin           : analog_signal;
              clamp_lo      : analog_v_level;
              clamp_hi      : analog_v_level)
                            RETURN analog_signal IS

VARIABLE   tau      : real;
VARIABLE   vout     : analog_signal;

BEGIN
   tau       := ( r.val * c.val);
   vout.val  := ( (pred1_vout.val * tau)    +
                  (vin.val * analog_time_delta_real) )
                 /(tau  + analog_time_delta_real);

-- clamp limits
   IF   (vout.val > clamp_hi.val) THEN vout.val := clamp_hi.val;
   ELSIF (vout.val < clamp_lo.val) THEN vout.val := clamp_lo.val;
   END IF;

   vout.kind := v;
   RETURN vout;

END rc;
```

Figure 5.   RC Low Pass Filter Function

The function parameter inputs are the R value, C value, the saturation limits passed into the amplifier model, the present value of the *vin* input signal, and the previous value of the output signal *vout*. The function returns the locally defined variable *vout* declared as type *analog_signal*. The other local variable *tau* is of type real and contains time information. The function first calculates the circuit time constant value *tau* based on the R and C values. Then the finite-difference is calculated, based on the result in Figure 1, generating the next value for *vout*. If the resulting value exceeds the saturation limits, *vout* is assigned the exceeded limit. The appropriate kind assignment is made and the aggregate value is returned via *vout*. Now that we have seen how the analog data structures shown in Figure 5 are used, their implementation details are described in Figure 6.

```
USE work.summing_bus_res_function.all;
PACKAGE analog_data_structure IS
..
.. analog time step
..
    CONSTANT analog_time_delta     : time := 100 ps;
    CONSTANT analog_time_delta_real: real := 0.1; ..ns
..
.. analog signal
..
    TYPE signal_mode IS (v, i);
    TYPE analog_signal IS RECORD
        val:   brf_real;
        kind:  signal_mode;
    END RECORD;
..
.. analog reference
..
    TYPE analog_reference_aref IS RECORD
        val:   real;
    END RECORD;
..
.. analog component
..
    TYPE component_element IS (res, cap, ind);
    TYPE analog_component IS RECORD
        val:   real;
        kind:  component_element;
    END RECORD;
..
.. analog threshold levels
..
    TYPE analog_v_level IS RECORD
        val:   real;
    END RECORD;
..
.. analog gain
..
    TYPE analog_vv_gain IS RECORD
        val:   real;
    END RECORD;
..
..general parms
..
    TYPE frequency IS RECORD
        val:   real;
    END RECORD;
    TYPE amplitude IS RECORD
        val:   real;
    END RECORD;
..
END analog_data_structure;
```

Figure 6.   Basic Analog Data Structure

Package *analog_data_structure* contains the analog time delta of VHDL physical type *time* used in all signal assignments for behaviors modeled with finite-difference techniques. It also has the same time delta of type *real* used in all functions evaluating these finite-difference equations. The second structure consists of a *record* type and is used to define analog signals. Signals of this type have one floating-point field containing the value of the signal and one enumeration field indicating its kind, either voltage or current. Physical types are not used for voltage and current values because a consistent set of units, as described in the previous chapter on transmission line modeling, cannot be maintained without resorting to scaling techniques. Similar record types are used for generic components. Generics of this type have one floating-point field containing the value of the generic while the other indicates its kind, either resistor, capacitor, or inductor. Finally, single field records describe the analog reference *aref*, the generic voltage saturation limits, the generic gain for the amplifier, and the frequency and amplitude for the function generator described later in applications.

Many analog circuits may be simulated with this basic analog data structure. However, to add the capability of current summing, the additional definition of a summing bus resolution function for determining resultant values of analog signals is required. Supporting the signal types of the previous *analog_data_structure* is the bus resolution function package shown in Figure 7. It is the same bus resolution function described in the previous chapter on transmission line modeling.

```
PACKAGE summing_bus_res_function IS
--
-- Declare subtype of a real for bus resolution function that is
-- an unconstrained array. The size of summing_val array equals the
-- number of signal assignment statements assigned to the common signal
-- that is currently calling the function.

   TYPE summing_val IS ARRAY ( integer RANGE<> ) OF real;
--
-- The function for bus resolution sums the currents or voltages at the
-- common signal (dot), hence, if one of the components is disabled its
-- signal assignment statement driving the signal should be assigned 0.0
-- in the behavior

   FUNCTION summing_brf_val ( data_array: summing_val )RETURN real;
--
   SUBTYPE brf_real IS summing_brf_val real;
--
END summing_bus_res_function;

PACKAGE BODY summing_bus_res_function IS
--

   FUNCTION summing_brf_val ( data_array: summing_val )RETURN real IS

      VARIABLE  i        : integer;
      VARIABLE  result   : real := 0.0;

   BEGIN
         FOR i IN data_array'RANGE
         LOOP
            result := result + data_array(i);
         END LOOP;

         RETURN result;
   END summing_brf_val;
--
END summing_bus_res_function;
```

Figure 7.   Basic Bus Resolution Function for Current Summing

It is important to note that this bus resolution function also supports the dotting of analog signals. The function simply sums the *brf_real* field for the drivers of analog signal assignments that share a common node and applies it to the signal.

Now that an analog circuit has been modeled as a behavior for simulation, the model must be verified against its corresponding physical implementation.

3.3 DESIGN VERIFICATION METHODOLOGY

To verify that the simulated behavior matches the physical interconnection, a two-step verification methodology similar to a hardware standard-cell approach is used. The first step verifies the correspondence between the physical design and the behavioral description at the cell level. This is the lowest level at which a VHDL behavior is written for a physical circuit. The second step simulates the interconnection of these verified cells along with concurrent signal assignments at a higher level such as the chip level.

3.3.1 Step 1: Cell Level

The VHDL capability of alternate architectures for an entity declaration is key to implementing the cell level design verification method depicted in Figure 8.

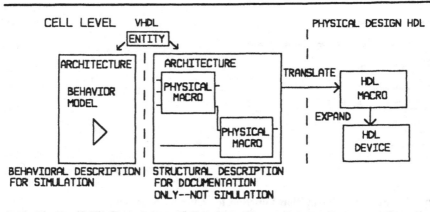

Figure 8. Step 1 -- Cell Design Verification

Two architectures are defined. The first models the behavior using the general simulation model described in Figure 2 while the second represents the structural design of the circuit. The structural architecture, unlike the behavior architecture, cannot be simulated. However, it does contain the information defining the physical macro or interconnection of the physical macros comprising the physical circuit of that cell. Here, a physical macro is a collection of devices and is the smallest physical unit that may be placed and wired in a hardware standard-cell design. The entity declaration, common to both the behavioral and structural architectures, ties the models together since the I/O to both architectures is common.

The first step in cell verification translates the structural architecture to an equivalent description in the existing physical design HDL. This macro-level HDL is then expanded into a device-level HDL such as SPICE or ASTAP. Next, the translated model is simulated on its simulator and the VHDL behavioral is simulated with the VHDL simulator. Each is driven in a similar fashion and the outputs are compared. If they agree to the level of detail desired, the "cell" verification is complete. The translated, physical design HDL description is then used to verify the layout data of the physical macros in the cell using an existing physical-to-logical verification methodology. For example, the translated SPICE deck with input stimulus for the amplifier example of Figure 3 is shown in Figure 9.

```
ANALYSIS OF SIMPLE 5V AMPLIFIER CIRCUIT
.WIDTH IN=72 OUT=80
.OPT LIMPTS = 2000  LVLCOD=2
.TEMP 27
.TRAN 200PS 300NS 0 200PS
*
****************************
* SIMPLE AMPLIFIER
****************************
* TEST CASE STIMULUS
VCC    10   0        5
VINP    7   0     SIN( 3.0,  0.1, 10MEG )
VINN    8   0     SIN( 3.0, -0.1, 10MEG )

IREF    0   9        1.020M
QR      9   9  11 0  NPN1
RR     11   0        0.250K

* LOAD
CL      6   0        15.9P

* AMPLIFIER CIRCUIT
R1     10   5        1.000K
R2     10   6        1.000K
R3      1   2        0.441K
R4      3   0        0.250K
R5      4   0        0.250K

Q1      5   7   1  0  NPN1
Q2      6   8   2  0  NPN1
Q3      1   9   3  0  NPN1
Q4      2   9   4  0  NPN1
****************************
*  MODELS
****************************
.MODEL NPN1 NPN IS=5.0E-17 BF=200 RE=1
****************************
*  OUTPUT
****************************
.PRINT TRAN  V(7) V(8) V(6)
.END
```

Figure 9. Translated SPICE Description for Amplifier

Once the structural architecture and its alternate behavioral architecture are verified at the "cell" level, that component may be referenced from any higher level architecture and the correspondence between the behavioral and structural architectures is correct by construction. With the cell-level verification defined, a basis is formed for chip-level verification.

3.3.2 Step 2: Chip Level

The VHDL capability of alternate configurations for components within an architecture is key to implementing the chip-level design verification method depicted in Figure 10.

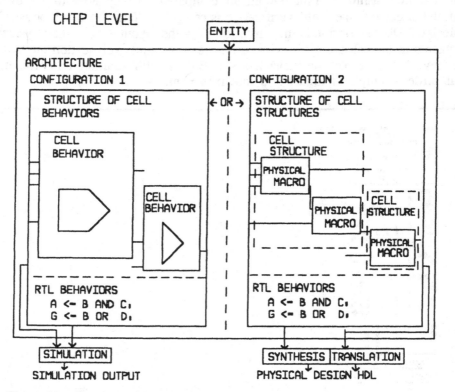

Figure 10.   Step 2 -- Chip Level Design Verification

The chip-level architecture consists of cell component references and RTL equations. After the design has been completed, two configurations for the chip architecture are defined. The first configures the instantiated components with their entity declarations and behavioral architectures forming a structure of cell behaviors. This results in a simulation model of the components referenced in the chip architecture. These component behaviors combined with RTL equations are simulated with the VHDL simulator. Design errors are removed by analyzing simulation results.

The second configuration configures the instantiated components with their entity declarations and structural architectures forming a structure of cell structures. This results in a structural model for the components referenced in the architecture which reflects the physical macro interconnections of the physical circuit. The components are then translated to a form that can be used to drive physical design tools. The RTL equations are synthesized into the same form. Combining the two provides a complete description of the chip in the physical design HDL which accurately reflects the VHDL behavioral chip description. This description in the physical design HDL is then used in placement and routing to build the chip.

3.4 APPLICATION OF ANALOG-DIGITAL BEHAVIORS

These basic techniques have been applied to a mixed analog-digital chip designed with the analog standard-cell approach presented in Reference [2]. This Bipolar/CMOS chip is 30.25mm square, has a 40%-60% analog-digital mix, and is used in hard disk drive signal processing applications. Functions which have been simulated include analog-to-digital converters, digital-to-analog converters, fixed gain amplifiers, variable gain amplifiers, envelope detectors, first and second order filters, voltage-to-current converters, and voltage controlled oscillators. The function and coarse feedback timing at the analog-digital interface were verified through simulation and the corresponding interconnection was checked. The test case stimulus used for verifying these functions was applied by VHDL "test equipment" models. As an example, the function generator model used to drive the amplifier is detailed.

3.4.1 Function Generator Model

The VHDL simulation model for the sine wave function generator shown in Figure 11 is a different type of analog model. Unlike the VHDL amplifier model, finite-difference equations are not used to calculate values. However, output values are still generated based on the *analog_time_delta*.

```
USE work.analog_data_structure.all;
USE work.scientific.all;
-- interface
ENTITY func_gen IS
     GENERIC ( amp      : amplitude     := ( val => 1.0 ) ;
               offset   : analog_v_level := ( val => 0.0) ;
               freq     : frequency     := ( val => 0.001) );
     PORT    ( v_sine  : OUT  analog_signal := (val => 0.0, kind => v);
               v_sine2 : OUT  analog_signal := (val => 0.0, kind => v));
END func_gen;

-- body
ARCHITECTURE func_gen_behavior OF func_gen IS
   BEGIN

P1: PROCESS
       VARIABLE temp_v, simtime : real := 0.0;
       CONSTANT period : real := 1.0/freq.val;
       BEGIN
       temp_v := amp.val * sin(6.2832*freq.val*simtime);

       v_sine.val <= temp_v + offset.val;
       v_sine2.val <= -temp_v + offset.val;

       WAIT FOR analog_time_delta;
       IF simtime <= period and
          simtime > (period - analog_time_delta_real) THEN
          simtime := analog_time_delta_real;
       ELSE
          simtime := simtime + analog_time_delta_real;
       END IF;

       END PROCESS;

END func_gen_behavior;
```

Figure 11. Basic Function Generator Model

The function generator model entity declaration contains generic and port declarations. The two output ports *v_sine* and its complement *v_sine2* drive voltage sine waves. Several VHDL generics customize the function generator. The sine wave amplitude, voltage offset and frequency are determined by *amp*, *offset* and *freq*; respectively.

The architectural body consists of one process statement, P1. First, the *period* for the sine wave is defined as the reciprocal of the frequency, *freq*. In the process statement, *temp_v* contains the AC output value. This value is summed and subtracted from the *offset* and assigned to signals *v_sine* and *v_sine2*, respectively. The process then waits for an *analog_time_delta* and increments *simtime* when the *period* has not been reached. If the period is reached, *simtime* is reset to *analog_time_delta_real*. This effectively provides a counter with modulus of 1/(freq * analog_time_delta_real). The only function referenced in this model is the sine function *sin* shown in Figure 12.

```
--
--funciton for sine
--
  FUNCTION sin( x: real ) RETURN real IS

    VARIABLE  sinx, x2, xtemp : real;
    TYPE const_array IS ARRAY (1 to 12) of real;
    CONSTANT fact_array : const_array := (6.0,
                                          120.0,           -- 3!
                                          5040.0,          -- 5!
                                          362880.0,        -- 7!
                                          39916800.0,      -- 9!
                                          6.22702e9,       -- 11!
                                          1.307674e12,     -- 13!
                                          3.556874e14,     -- 15!
                                          1.21645e17,      -- 17!
                                          5.109094e19,     -- 19!
                                          2.585201e22,     -- 21!
                                          1.551121e25);    -- 23!
  BEGIN                                                    -- 25!
      x2 := x**2;
      xtemp := x;
      sinx := x;
      FOR i IN fact_array'RANGE LOOP
        xtemp := xtemp * x2;
        IF (i rem 2) = 1 THEN
          sinx := sinx - (xtemp / fact_array(i));
        ELSE
          sinx := sinx + (xtemp / fact_array(i));
        END IF;
      END LOOP;
      RETURN sinx;
  END sin;
--
```

Figure 12.   Function Generator Sine Function

This function approximates the sine function by using the first thirteen terms of its Taylor series expansion. The *fact_array* table provides the factorials for these terms so recalculating the factorials is not necessary. This reduces the CPU time required to execute the function.

3.4.2 Receiver Model

The receiver model used is a simplified version of the transmission line receiver model presented in the previous chapter. The receiver input node type is changed to type *analog_signal* and the mode to IN. Also, the VHDL generic *r_input* is removed implying an ideal infinite input resistance. The resulting receiver model is shown in Figure 13.

Dynamically, the receiver exhibits the same hysteresis switching characteristics. The output switches to a logic "1" state after a delay when the output is in a logic "0" state and the input voltage exceeds the higher threshold. Conversely, the output switches to a logic "0" state after a delay when the output is in a logic "1" state and the input voltage falls below the lower threshold. Finally, the output remains in the same state if neither condition is satisfied.

```
USE work.analog_data_structure.all;

-- interface
ENTITY receiver1 IS
    GENERIC ( v_sw_hi  : real   := 1.0 ; -- volts
              v_sw_low : real   := 1.0 ; -- volts
              t_delay  : time   := 1 ns        );
    PORT    ( rec_in   : IN   analog_signal := (val=> 0.0, kind=> v);
              rec_out: BUFFER bit         );
END receiver1;

-- body
ARCHITECTURE receiver1_behavior OF receiver1 IS
    BEGIN
    rec_out <= '1' AFTER t_delay WHEN (rec_in.val > v_sw_hi ) ELSE
               '0' AFTER t_delay WHEN (rec_in.val < v_sw_low) ELSE
        rec_out AFTER t_delay;

--
END receiver1_behavior;
```

Figure 13.   Simplified Hysteresis Reciever Model

3.4.3 Network Example

To illustrate analog-digital simulation the network shown in Figure 14 combines the function generator (FG), amplifier (A), and the hysteresis receiver (R).

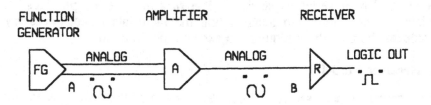

Figure 14.   Analog Network Example

Each model is implemented as a VHDL entity. Specifically, the function generator drives the amplifier (A) differentially at node A. The amplifier reads the voltage at node A, amplifies and phase shifts it, then drives node B. The hysteresis receiver reads the voltage at node B and drives a logic output after a time delay.

The salient part of the VHDL model for this analog-digital simulation example is shown in Figure 15.

```
USE work.analog_data_structure.all;
..
ENTITY main_amp IS
END main_amp;
..
ARCHITECTURE main_amp_sample OF main_amp IS
    SIGNAL   vinp, vinn    : analog_signal := (val => 0.0, kind => v);
    SIGNAL   vout          : analog_signal := (val => 0.0, kind => v);
    SIGNAL   aref          : analog_reference_aref := (val => 1.2);
    SIGNAL   recl_logic_out : bit := '0';
-- Partial Declarative Region
    ..
  BEGIN
    ..

  FG1 : func_gen GENERIC MAP ( (val => 0.1  ),
                               (val => 3.0  ),
                               (val => 0.010) )
               PORT MAP   (vinp, vinn);

  AMP1: amplifier GENERIC MAP ( (val =>    4.0),
                                (val =>    4.0),
                                (val =>    3.0),
                                (val =>    5.0),
                                (val =>    1.0, kind => res),
                                (val =>   15.9, kind => cap) )
               PORT MAP   ( vinp, vinn, aref, vout);

  R1: receiver1   GENERIC MAP (  v_sw_hi  => 4.1,
                                 v_sw_low => 3.9,
                                 t_delay  => 1 ns    )
               PORT MAP   ( vout, recl_logic_out );
  ..
END main_amp_sample;
```

Figure 15. Amplifier Network Entity

The component instantiations correspond directly to Figure 14. Function generator FG1 drives the true and complement of a 0.1 V peak sine wave at 0.010 gigahertz centered on 3.0 volts DC. To match the circuit behavior, the VHDL amplifier model was passed a gain of 4.0, an output common-mode level of 4.0 V, an R value of 1 Kohm, a C value of 15.9 pf, and saturation limits of 3.0 V and 5.0 V. Finally, the receiver was characterized by a high switch level of 4.1 V, a low switch level of 3.9 V, and a delay of 1 ns.

3.4.4 Simulation Results

For comparison, the resulting waveforms from both the VHDL behavioral simulation and SPICE circuit simulation are shown in Figure 16. The first three $V(6)$, $V(7)$, and $V(8)$ represent the SPICE results for the amplifier inputs and output based on the source deck shown in Figure 9.

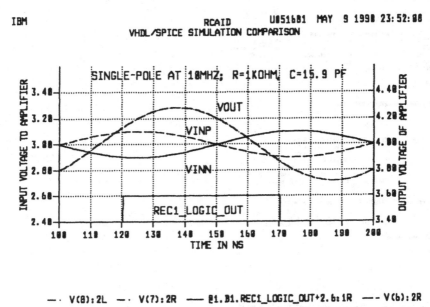

IBM   RCAID   U051681 MAY 9 1990 23:52:00
                     VHDL/SPICE SIMULATION COMPARISON

Figure 16.   Plot Showing Amplifier Driven at -3dB Frequency

It is difficult to discern the SPICE waveforms from the VHDL waveforms because of the excellent agreement in simulation results. Both the VHDL and SPICE amplifier models, driven by a 400 mV peak-to-peak differential sine wave at the -3db frequency of 10 MHZ, showed a difference in gain and phase for the single-ended outputs of less than 1.5%. However, comparing the VHDL simulation result for the single-ended output to the predicted ideal behavioral result (a 566 mV peak-to-peak sine wave centered on 4.0 V and delayed by 45 degrees with respect to the differential input signal) revealed a gain and phase error less than 0.5%.

Furthermore, the receiver logic output, *rec1_logic_out,* for VHDL simulation is also shown in Figure 16. Its logic value is offset so it appears on the figure with a logic value of '1' corresponding to 3.6. Note that it switches high as the voltage input exceeds the high switching threshold of 4.1 V near 120 ns and switches low as the amplifier output voltage falls below 3.9 V near 170 ns.

This type of simulation and modeling has been applied to the other analog-digital behaviors, described earlier for the Bipolar/CMOS chip, yielding similar results.

3.4.5 Limitations

The general simulation model presented was applied to transfer function modeling. These transfer function simulation models are limited to realizable analog functional blocks. A general transfer function model for a resistor, capacitor, or inductor is not possible; however, passive networks comprised of these elements may be modeled if the desired transfer function is defined.

Since a block's output(s) are a function of its input(s), each analog behavior must account for the loading on its output. Hence, circuits such as an analog switch which present large impedance changes to a block output during simulation cannot be modeled unless bidirectional connections are used. VHDL can describe bidirectional (INOUT) ports, hence the load may broadcast its value to the behavior at time zero and update it during simulation providing a method of coupling the functional block to its load. These methods, which were used in the previous chapter, reduce simulation efficiency and increase model complexity.

To summarize, transfer function modeling is most applicable to circuits with well defined transfer functions and low sensitivity to loading variation for the specific application [5].

3.4.6 Model Usage

These models primarily supplement logic models for mixed analog-digital simulation. The additional aspects modeled by the analog behavior models enable combined VHDL analog-digital simulation.

The added simulation capability when using these models does increase CPU time for a simulation and may pose problems for synthesis. The accuracy/runtime trade-off is directly related to the *analog_time_delta* used. Specifying a delta of 100 ps typically uses the same order of magnitude CPU time as the equivalent SPICE simulation. Execution time increases ten times for 10 ps resolution and decreases 10 times for 1 ns resolution. Typically,

100 ps resolution is adequate for analog hardware with time constants longer than 2 ns and input frequencies below 100 MHZ.

Careful examination of the time constants and input signal frequency content for the analog circuits must be made before the time delta is selected. Since the ratios of the time delta to circuit time constants are used in finite-difference calculations, circuit time constants must be much greater than the time delta to obtain meaningful results. Also, the input signal harmonic content should be much less than the reciprocal of the analog time delta. Again, in using finite-difference equations a small variation in input signals between successive time delta steps reduces simulation error.

Finally, care must be taken in verifying the correspondence between the VHDL analog simulation model and the device-level circuit it represents. Multiple simulations in SPICE or ASTAP and in VHDL are required to characterize the VHDL model for a circuit. Also, since it is difficult to model all aspects of a circuit in a simple VHDL simulation, assertions should be used to verify model operation within model limitations.

3.5 SUMMARY

VHDL is an effective modeling language for analog-digital behavioral simulation and design verification. VHDL can describe analog and digital behaviors by using the general simulation model. Coupling the present modeling techniques for digital circuits to the behavioral modeling techniques described earlier for analog circuits establishes the basis for combined analog-digital behavioral simulation. The standard-cell approach to design verification can compare the structural models reflecting macro-level physical design to their behavioral counterparts at the cell level. This establishes the basis for chip-level description and design verification. At the chip level, the VHDL capability of alternate configurations allows the designer to expand the components of the model hierarchy behaviorally or structurally. Any RTL behaviors are common to both configurations. They are not modified for simulation, but for structural expansion they are synthesized into a macro-level format for the physical design tools.

Applying the described modeling techniques for simulation and design verification yielded excellent results. Expanding on the techniques and general models presented, a standard-cell analog-digital chip has been simulated and verified. This chip simulation was possible since the analog circuits were modeled with well defined transfer functions and exhibited low sensitivity to loading variation in this application. Not all analog-digital designs may be suitably modeled in this manner so the applicability and

limitations of transfer function modeling described earlier should be considered for each analog-digital design. Developing additional applications
of the general simulation model and the full potential of VHDL analog-
digital event-driven simulation requires further research. With respect to
design verification, as is the case of any standard-cell approach, the possibility of insufficient test coverage, behavioral detail, and inadequate modeling for varying loading conditions exists. With reasonable care their
associated problems are minimized and are outweighed by the benefits of
simulating the analog-digital interaction behaviorally and verifying the
analog-digital interconnection. VHDL, rich in the constructs found in both
an HDL and a programming language, has the flexibility and structure necessary for an effective analog-digital behavioral simulation language reflecting the analog-digital interconnection.

REFERENCES

[1] "VHDL User's Manual," No. IR-MD-065-1, Intermetrics Inc., Bethesda,
MD, 1985

[2] M. A. Masetti, et al., "A CMOS-Based Analog Standard Cell Product
Family," IEEE J. Solid-State Circuits, Vol. SC-24, No. 2, pp. 370-379, Apr.
1989.

[3] J. A. Cadzow, and H. R. Martens, Discrete-Time and Computer Control
Systems. Prentice-Hall: Englewood Cliffs, NJ, 1970.

[4] B. R. Stanisic and M. W. Brown, "VHDL Modeling for Analog-Digital
Hardware Designs," in Proc. IEEE Int. Conf. Computer-Aided Design, pp.
184-187, Nov. 1989.

[5] R. A. Cottrell, "Event-Driven Behavioural Simulation of Analogue
Transfer Functions," in Proc. European Design Automation Conference,
Glasgow, U. K., Mar. 1990.

[6] J. M. Schoen, "Analog Hardware Descriptive Language IEEE Standards
Committee 30 Report," VHDL Users' Group Fall 1989 Meeting, pp. 12.01
- 12.03, Redondo Beach, CA, Oct. 1989.

[7] B. R. Stanisic, "VHDL Simulation Modeling for a Phase-Locked Loop,"
VHDL Users' Group Spring 1990 Meeting, pp. 7.01 - 7.13, Boston, MA,
Apr. 1990.

[9] T. Corman and M. Wimbrow, "Coupling a Digital Logic Simulator and an Analog Circuit Simulator," VLSI Systems Design, CMP Publications Inc., pp. 38-47, Feb. 1988.

[10] B. O'Rouke, "Behavioral Modeling of Digital Devices in an Analog Simulation Environment," VLSI Systems Design, CMP Publications Inc., pp. 16-25, Jan. 1988.

[11] D. Thelen and J. MacDonald, "Simulating Mixed Analog-Digital Circuits on a Digital Simulator," in Proc. IEEE Int. Conf. Computer-Aided Design, pp. 254-257, Nov. 1988.

[12] A D&T Roundtable, "Mixed-Mode Simulation," IEEE Design & Test, pp. 67-75, Feb. 1989.

[13] B. R. Stanisic, "IEEE SCC 30 VHDL Analog Application Working Group," in Proc. IEEE Workshop on Analog Circuit Engineering, Sec. 3.4, Jan. 1990.

[14] C. Visweswariah and R. A. Rohrer, "Piecewise Approximate Circuit Simulation," in Proc. IEEE Int. Conf. Computer-Aided Design, pp. 248-251, Nov. 1989.

[15] "IEEE Standard VHDL Language Reference Manual," IEEE Standard 1076-1987, New York, NY.

[16] E. Kreyzig, Advanced Engineering Mathematics. John Wiley & Sons: New York, NY, 1972.

Chapter 4

Modeling of
Analog-Digital Loops
in VHDL

Balsha R. Stanisic

International Business Machines
Rochester, Minnesota

The previous chapters introduced the engineer to some of the basic concepts in using VHDL for behavioral modeling. This chapter is written to assist the applications engineer familiar with Automatic Gain Control (AGC) and Phase-Locked (PL) loops in applying those concepts. The intention is to incorporate loop behavior into analog-digital modeling and demonstrate the resulting analog-digital simulation. Described is an approach for modeling combined analog-digital loop behavior -- centered on the VHDL behavioral models used to simulate the AGC and PL loops. This approach enables the VHDL simulation of mixed analog-digital hardware designs incorporating these control loops.

4.1 INTRODUCTION

These models use the behavior modeling approach described in the previous chapter. The behaviors implementing components of the loops use the underlying concepts, data structures, and general model structure previously described. Details of all behaviors for each loop introduced here are provided and a description of each implementation in VHDL is given. Finally, a network example and simulation results are shown for each type of loop. The loops are described in separate sections beginning with the AGC loop.

4.2 AGC LOOP BEHAVIORAL MODELS

The basic behavioral models required to simulate the AGC loop are the variable gain amplifier (VGA), the envelope detector (ED), and the integrating capacitor (C) shown in Figure 1.

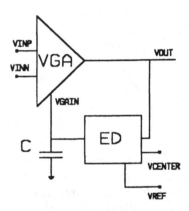

Figure 1. Basic AGC Loop Simulation Models

Each behavioral model is implemented with a VHDL entity. Each entity implements the behavior with a well defined transfer function and is interconnected with signals of type *analog_signal*. Specifically, the VGA gain is controlled by the voltage at the gain node (VGAIN) and drives a voltage output (VOUT) based on its differential input (VINP - VINN). The envelope detector reads this output voltage and compares it to the center voltage (VCENTER) and reference voltage (VREF) to set the envelope amplitude. This comparison results in a current output being driven to the integrating capacitor (C) to adjust the gain. Finally, the integrating capacitor integrates the current and drives a voltage at the gain control node (VGAIN) of the VGA, closing the loop.

4.2.1 Variable Gain Amplifier (VGA)

The VGA is the most important element of the loop. It is the variable gain element driving the output voltage. It calculates the proper gain based on gain voltage (VGAIN). The VGA symbol and its gain -vs- VGAIN voltage transfer curve are depicted in Figure 2.

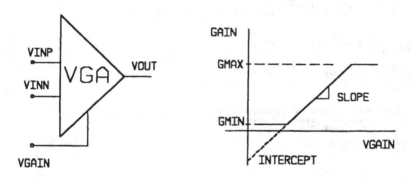

Figure 2.   Variable Gain Amplifier Symbol and Transfer Curve

The VGA exhibits several static characteristics. First, the VGA gain varies between the gmin and gmax value as a function of the voltage on the gain control input (VGAIN). In this case, the VGA has a linear gain -vs- VGAIN voltage transfer curve with a near-zero value of gain at low control voltages (VGAIN). Aside from the variable gain aspect, the amplifier is similar to an ordinary differential to single-ended amplifier described in the previous analog chapter.

Dynamically, the VGA gain varies with the gain voltage (VGAIN). A maximum slew rate is not incorporated into the model; however, a single pole roll-off is modeled in the data path.

The VHDL model for the variable gain amplifier (VGA) is shown in Figure 3.

```
USE work.analog_data_structure.all;
USE work.analog_passive_behaviors.all;
USE work.analog_active_behaviors.all;

-- interface
ENTITY vg_amplifier  IS
      GENERIC ( slope       : real;
                intcpt      : analog_vv_gain;
                vcmo        : analog_v_level;
                vout_min    : analog_v_level;
                vout_max    : analog_v_level;
                r           : analog_component;
                c           : analog_component);
      PORT    ( vinp    : IN      analog_signal;
                vinn    : IN      analog_signal;
                vgain   : IN      analog_signal;
                vout    : BUFFER analog_signal := (val => 0.0, kind => v));

BEGIN
      ASSERT( (r.val > 0.0) AND (c.val > 0.0) )
          REPORT "ERROR: R and C must be > 0"
          SEVERITY failure;

      ASSERT( vout_max.val > vout_min.val )
          REPORT "ERROR: vout_max must be > vout_min"
          SEVERITY failure;
END vg_amplifier;

-- body
ARCHITECTURE vg_amplifier_behavior OF vg_amplifier IS
      SIGNAL ac_in, ideal_vout, pvout : analog_signal :=
                (val => 0.0, kind => v );

      SIGNAL gain     : analog_v_level :=
                (val => 0.0 );

-- instansiate
   BEGIN
      -- Ideal Amplifier
      gain   <= variable_gain( vgain, slope, intcpt );
      ac_in.val   <= (vinp.val - vinn.val) * 0.5;
      ideal_vout.val <= vcmo.val + ( ac_in.val *  gain.val );

      -- Single Pole
      vout      <= rc(r, c, pvout, ideal_vout, vout_min,vout_max);
      pvout     <= vout AFTER analog_time_delta;

END vg_amplifier_behavior;
```

Figure 3.   Variable Gain Amplifier Model

The VGA entity declaration is similar to the amplifier model described in the last chapter. Again, the VHDL generics customize the RC dominant pole roll-off point, the output common-mode voltage, and the saturation limits; however, the generics *slope* and *intcpt* are added for describing the gain/voltage transfer curve facilitating the variable gain aspect of the amplifier.

In the VGA architectural body, the model first calculates the gain based on the control voltage input, *vgain*, and slope/intercept generic values using the *variable_gain* function. Then it proceeds just like the amplifier example of the previous chapter. The details of the *variable_gain* function are shown in Figure 4.

```
--
--
-- function for VGA - returns a gain for the given voltage
--
   FUNCTION variable_gain ( vin : analog_signal;
                            slope     : real; --ghz/v
                            intercept : analog_vv_gain)
                                 RETURN analog_v_level IS

   VARIABLE   gain       : analog_v_level;

   BEGIN

     gain.val := (slope * vin.val) + intercept.val;

     IF (gain.val < 0.001) THEN
        gain.val :=  0.0;
     END IF;

     IF (gain.val > 1000.0) THEN
        gain.val := 1000.0;
     END IF;

     RETURN gain;
   END variable_gain;
--
```

Figure 4.   Variable_gain Function for Variable Gain Amplifier

For the VGA depicted in Figure 2, the *variable_gain* function first calculates *gain* based on *slope*, *intercept*, and gain voltage *vin*. Then for negative or near zero values of gain, a minimum gain value of zero is assigned. Similarly, for gain values greater than 1000, a maximum gain of 1000 is assigned. Finally, the *gain* is returned. The VGA output drives the next behavior in the loop, the envelope detector.

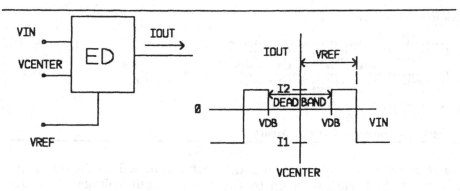

Figure 5.   Envelope Detector Transfer Curve and Block Diagram

4.2.2 Envelope Detector (ED)

The envelope detector depicted in Figure 5 controls the loop gain by varying its current output over three regions of operation. For input signal voltages (VIN) near the center voltage (VCENTER) a deadband is defined with no current flow. For voltages outside the center voltage (VCENTER) plus or minus the reference voltage (VREF) the envelope detector sinks a large current (I1). Finally, for voltage between this region and the deadband cutoff, some current is sourced (I2). The detector presents an infinite input resistance to the amplifier, has delay from input to output, and the output current is typically integrated on a capacitor.

Dynamically, the detector output switches to sourcing some current (I2) to increase gain when the input voltage (VIN) exceeds the deadband threshold (VDB) and not the reference voltage (VREF). The output switches to sinking larger current (I1) to reduce gain when the input voltage (VIN) exceeds the reference voltage (VREF). Finally, the output switches to a zero current state when the input voltage approaches the center reference voltage (VCENTER).

The VHDL model for the envelope detector is shown in Figure 6.

```
USE work.analog_data_structure.all;
-- interface
ENTITY env IS
     GENERIC ( imag_sink   : real := 0.5 ;
               imag_source : real := 0.35 ;
               db_percent  : real := 0.2 ;
               t_delay     : time := 1 ns );
     PORT    ( vin     : IN      analog_signal;
               iout    : BUFFER analog_signal := (val => 0.0, kind => 1);
               vref    : IN      analog_signal;
               vcenter : IN      analog_signal );
END env;

-- body
ARCHITECTURE env_behavior OF env IS
     BEGIN
     iout.kind <= 1;
     iout.val  <=     -imag_sink   AFTER t_delay WHEN
     (   (vref.val)       < (vin.val - vcenter.val) OR
         (vref.val)       < (vcenter.val - vin.val)      )       ELSE
                     imag_source AFTER t_delay WHEN
     (   (vref.val)*(db_percent) < (vin.val - vcenter.val) OR
         (vref.val)*(db_percent) < (vcenter.val - vin.val)      )       ELSE
                     0.0    AFTER t_delay;
--
END env_behavior;
```

Figure 6.   Envelope Detector Model

The envelope detector model entity declaration defines the interface. The input port, *vin*, is driven by the VGA output voltage. The detector output port, *iout*, drives a current to the integrating capacitor. The other two input ports, *vref* and *vcenter*, provide the reference voltages for envelope amplitude and center voltage, respectively. The VHDL generics *imag_sink* and *imag_source* customize the magnitude of the discharging and charging integration currents (IOUT). Also, the generic *db_percent* defines the deadband thresholds (VDB) as a percentage of *vref*.

The architectural body implements the behavior. This detector consists of one conditional signal assignment for determining the output current (IOUT). It defines the three regions of operation and applies a fixed value of current to *iout* for each region. Note that the deadband region is VCENTER plus or minus (DB_PERCENT * VREF). The current driven in the other two regions is integrated by the last behavioral model in the loop, the capacitor.

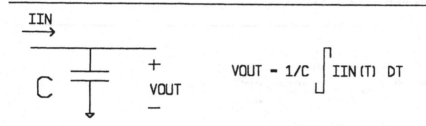

Figure 7. Integrating Capacitor Block Diagram and I-V Relation

4.2.3 Integrating Capacitor (C)

The capacitor filter shown in Figure 7 is a simple one capacitor integrator. The output voltage (VOUT) is the integral of the input current (IIN). The output voltage increases for current into the capacitor and decreases for current out of the capacitor.

The VHDL model for this capacitor is shown in Figure 8.

```
USE work.analog_data_structure.all;
USE work.analog_passive_behaviors.all;
-- interface
ENTITY cap_filter IS
    GENERIC ( c : analog_component := (val => 0.0, kind => cap ) );
    PORT    ( iin    : IN      analog_signal;
              vout   : BUFFER analog_signal := (val => 0.0, kind => v));

BEGIN
    ASSERT( c.val > 0.0 )
        REPORT "ERROR: C must be > 0"
        SEVERITY failure;

END cap_filter;

-- body
ARCHITECTURE cap_filter_behavior OF cap_filter IS
    SIGNAL pvout : analog_signal :=
            ( val => 2.7, kind => v );

    BEGIN
      vout  <= capf(c, iin, pvout );
      pvout <= vout AFTER analog_time_delta;

END cap_filter_behavior;
```

Figure 8. Integrating Capacitor Model

The integrating capacitor model is derived much like the single pole example described in the previous chapter. In the entity declaration, the current input, *iin*, and voltage output, *vout*, are defined along with the VHDL generic for the integrating capacitor value, *c*. In the architectural body, the output voltage is initialized to a reasonable starting value and the finite-difference equation is calculated with the function described in Figure 9.

```
--
--
-- function for Cap filter- integrator      vout-C-gnd
--
   FUNCTION capf ( c          : analog_component;
                   iin        : analog_signal;
                   predl_vout: analog_signal )
                           RETURN analog_signal IS

   VARIABLE   vout  : analog_signal;

   BEGIN
         vout.val := ( (iin.val/c.val)*analog_time_delta_real +
                                       predl_vout.val );

         vout.kind := v;
         RETURN vout;
   END capf;
```

Figure 9. Capf Function for Integrating Capacitor

The *capf* function calculates the next value of voltage based on the previous voltage value, *predl_vout*, and the present current value, *iin*. Then, it returns the output voltage value, *vout*, that drives the VGA gain control node, closing the loop.

4.3 APPLICATION OF AUTOMATIC GAIN CONTROL LOOP

AGC loops can be simulated using these models. Changing parameters is easily accomplished by modifying VHDL generics. The following network example illustrates one application.

4.3.1 Network Example

These models form a basis for simulating AGC loops in mixed analog-digital chip designs. An example network using these models is shown in Figure 10.

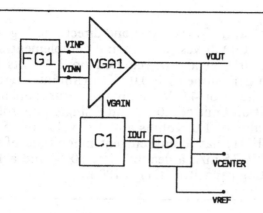

Figure 10.   AGC Loop Example

The variable gain amplifier (VGA1), envelope detector (ED1) and integrating capacitor (C1) are used. The salient part of the VHDL Model for this simulation example is shown in Figure 11. It includes the function generator (FG1) described in the previous chapter.

```
USE work.analog_data_structure.all;
--
ENTITY main_agc IS
END main_agc;
--
ARCHITECTURE main_agc_sample OF main_agc IS
    SIGNAL    vinp, vinn    : analog_signal := (val => 0.0, kind => v);
    SIGNAL    vout          : analog_signal := (val => 0.0, kind => v);
    SIGNAL    vgain         : analog_signal := (val => 0.0, kind => v);
    SIGNAL    vref          : analog_signal := (val => 0.1, kind => v);
    SIGNAL    vcenter       : analog_signal := (val => 4.0, kind => v);
    SIGNAL    iout          : analog_signal := (val => 0.0, kind => i);
-- Partial Declarative Region
    --
  BEGIN
    --

    FG1 : func_gen      GENERIC MAP ( (val => 0.01 ),
                                      (val => 3.0  ),
                                      (val => 0.010) )
                        PORT MAP (vinp, vinn);

    vga1: vg_amplifier GENERIC MAP (        100.0,    -- gain/v
                                      (val => -200.0),
                                      (val =>    4.0),
                                      (val =>    3.0),
                                      (val =>    5.0),
                                      (val =>    1.0, kind => res),
                                      (val =>   15.9, kind => cap) )
                        PORT MAP   ( vinp, vinn, vgain, vout);

    ed1: env            GENERIC MAP ( 0.5, 0.35, 0.2, 1 ns )
                        PORT MAP   ( vout, iout, vref, vcenter);

    c1: cap_filter      GENERIC MAP ( (val =>  100.0, kind => cap)  )
                        PORT MAP   ( iout, vgain);

--
END main_agc_sample;
```

Figure 11.   AGC Network Entity

The component instantiations correspond directly to Figure 10. The function generator (FG1) drives the true and complement of a 0.01 V peak sine wave at 0.010 gigahertz centered on 3.0 volts DC. The VGA (VGA1) has a gain slope of 100.0 /V, a gain intercept of -200.0, a common-mode voltage of 4.0 V, minimum saturation limit of 3.0 V, maximum saturation limit of 5.0 V, and a single pole roll-off characterized by an R value of 1.0 Kohms and a C value of 15.9 pf. The envelope detector (ED1) has a sink current magnitude of 0.5 mA, a source magnitude of 0.35 mA, a deadband of 20%, and a 1 ns delay. Finally, the integrating capacitor (C1) is 100 pf.

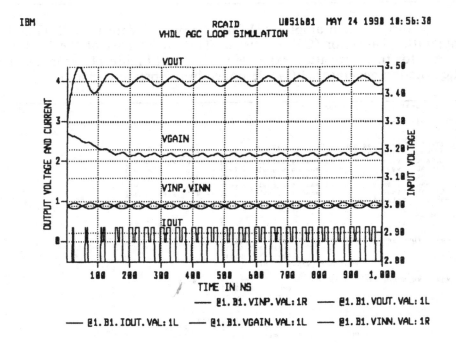

Figure 12. AGC Loop Example Waveforms

4.3.2 Simulation Results

The resulting waveforms from the VHDL simulation are shown in Figure 12. Note that these waveforms accurately reflect the theoretical behavior expressed in the AGC loop behavioral models.

4.4 PHASE-LOCKED LOOP BEHAVIORAL MODELS

The basic behavioral models required to simulate the Phase-Locked (PL) loop are the single shot (SS), phase/frequency detector (PD), charge pump (CP), second-order filter (F), and voltage controlled oscillators (VCO) shown in Figure 13.

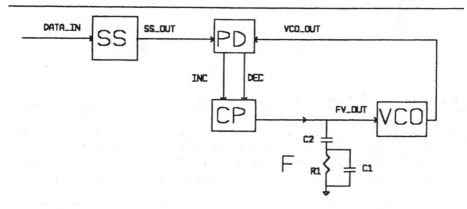

Figure 13.   Basic PLL Loop Simulation Models

Each behavior-level model has a well defined transfer function and is interconnected with signals of type *analog_signal* and *bit*. Specifically, the single shot output (SS_OUT) drives a fixed pulse width logic output for a logic input rising edge. At the target frequency, this results in a square wave with a 50% duty cycle. The phase/frequency detector (PD) compares this logic input (SS_OUT) with the voltage controlled oscillator logic output (VCO_OUT). Then, it outputs increment (INC) and decrement (DEC) digital pulses of width equal to the rising edge time difference between the two logic signals. The charge pump then sources an output current to the filter for increment pulses (INC) and sinks current from the filter for decrement pulses (DEC). The filter integrates the current and responds instantaneously since it is comprised of both capacitors (C1 & C2) and a resistor (R1). This filter voltage (FV_OUT) then drives the voltage controlled oscillator. The VCO adjusts the frequency of its logic output (VCO_OUT) based on this voltage input (FV_OUT), closing the loop.

4.4.1 Single Shot (SS)

The single shot is the only element in series with the data signal (DATA_IN) and is shown in Figure 14.

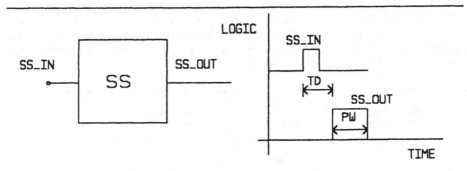

Figure 14. Single Shot Block and Timing Diagrams

The single shot exhibits several static characteristics. First, it is an edge sensitive digital behavior which outputs a fixed pulse width (PW) logic signal (SS_OUT). It has delay (TD) from input to output and is often used for duty cycling narrow pulses of data input (SS_IN).

```
-- interface
ENTITY single_shot IS
      GENERIC (    pulse_width: time := 50 ns;
                   t_delay     : time := 1 ns );
      PORT    (    ss_in  : IN     bit;
                   ss_out : BUFFER bit);

BEGIN
      ASSERT ( pulse_width >= 1 ns )
        REPORT "ERROR: pulse width of single shot must be >= 1 ns"
        SEVERITY failure;

      ASSERT ( t_delay >= 0 ps )
        REPORT "ERROR: time delay through single shot must be >= 0 ps"
        SEVERITY failure;

END single_shot;

-- body
ARCHITECTURE single_shot_behavior OF single_shot IS
  BEGIN
B1: BLOCK(ss_in = '1' AND ( NOT ss_in'STABLE) )
    BEGIN
        ss_out <= GUARDED '1' AFTER t_delay, '0' AFTER
                                    (pulse_width + t_delay);
    END BLOCK b1;
END single_shot_behavior;
```

Figure 15. Single Shot Model

The VHDL model for the single shot is shown in Figure 15.

The single shot entity declaration defines two ports, *ss_in* and *ss_out* and VHDL generics *pulse_width* and *t_delay*. The assertion checks for valid delay and pulse width.

The single shot architectural body consists of one block sensitive to a rising edge on the data input *ss_in*. The signal assignment contained in the block outputs a logic "1" after the time delay, *t_delay*, and a logic "0" after the pulse width, *pulse_width*, plus a time delay. This logic output drives one input to the phase/frequency detector described next.

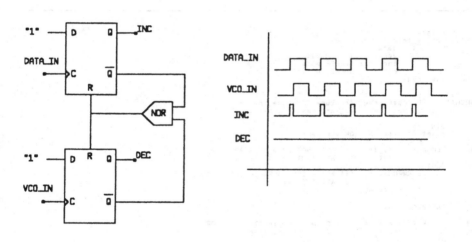

Figure 16.   Phase/Frequency Detector Block and Timing Diagrams

4.4.2 Phase/Frequency Detector (PD)

The phase/frequency detector depicted in Figure 16 exhibits several static characteristics. The detector compares two edges of a data (DATA_IN) and oscillator (VCO_IN) logic signal. When the oscillator edge (VCO_IN) lags the data edge (DATA_IN) an increment pulse (INC) is output to increase the frequency of the oscillator. When the oscillator edge (VCO_IN) leads the data edge (DATA_IN) a decrement pulse (DEC) is output to decrease the frequency of the oscillator. The detector detects 360 degrees so it does not "slip" cycles. The detector has delay from input to output and is strictly a logic behavior.

Dynamically, the detector increment output (INC) switches to a logic "1" when the data input (DATA_IN) switches high before the VCO input (VCO_IN) switches high. Then, the increment output (INC) switches to a logic "0" when the VCO input (VCO_IN) switches high. Conversely, the decrement output (DEC) switches to a logic "1" when the VCO input (VCO_IN) switches high before the data input (DATA_IN) switches high. Finally, the decrement output (DEC) switches to a logic "0" when the data input (DATA_IN) switches high.

The VHDL model for the phase/frequency detector is shown in Figure 17.

```
-- interface
ENTITY phase_freq IS
    PORT    (    data_in: IN      bit;
                 vco_in : IN      bit;
                 inc    : BUFFER bit;
                 dec    : BUFFER bit);
END phase_freq;

-- body
ARCHITECTURE phase_freq_structure OF phase_freq IS
      SIGNAL data_bar, vco_bar, reset  : bit := '0';

    BEGIN
      reset <= data_bar NOR vco_bar;

BA:    BLOCK ( (data_in = '1' AND (NOT data_in'STABLE) ) OR
              (reset   = '1' AND (NOT reset'STABLE) )   )
      BEGIN
         inc <= GUARDED  '1' AFTER 1 ns WHEN (reset = '0') ELSE
                         '0' AFTER 1 ns;
         data_bar <= NOT inc;

      END BLOCK BA;

BB:    BLOCK ( (vco_in = '1' AND (NOT vco_in'STABLE) ) OR
              (reset   = '1' AND (NOT reset'STABLE) )   )
      BEGIN
         dec <= GUARDED  '1' AFTER 1 ns WHEN (reset  = '0') ELSE
                         '0' AFTER 1 ns;
         vco_bar <= NOT dec;

      END BLOCK BB;

END phase_freq_structure;
```

Figure 17.   Phase/Frequency Detector Model

The phase/frequency detector model entity declaration contains four logic ports: *data_in*, *vco_in*, *inc*, and *dec*.

The architectural body implements the logic diagram depicted in Figure 16. Two blocks, *BA* and *BB*, model the two D edge triggered flip-flops used in the phase detector. The asynchronous reset for the latches is controlled by the signal assignment statement implementing the NOR function. The true outputs of the flip-flops drive the next behavior in the loop, the charge pump.

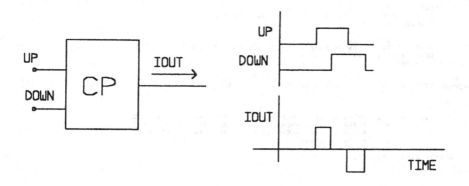

Figure 18.   Charge Pump Block and Timing Diagrams

4.4.3 Charge Pump (CP)

The charge pump shown in Figure 18 sources and sinks current based on the increment (UP) and decrement (DOWN) inputs. When the increment (UP) goes high, the charge pump sources current to the filter. When the decrement (DOWN) goes high, the charge pump sinks current from the filter. When both increment (UP) and decrement (DOWN) are in the same state, no current is driven. The charge pump exhibits delay from input to output and is an analog-digital behavior.

Dynamically, the charge pump output (IOUT) sources current when the increment input (UP) switches to a logic "1" provided the decrement input (DOWN) is logic "0". Conversely, the output (IOUT) sinks current when the decrement input (DOWN) switches to a logic "1" provided the increment input (INC) is low.

The VHDL model for the charge pump is shown in Figure 19.

```
USE work.analog_data_structure.all;

-- interface
ENTITY charge_pump IS
    GENERIC (    imag   : real := 0.5;    --current magnitude in ma
                 t_delay: time := 1 ns ); --time delay
    PORT    (    up     : IN  bit;
                 down   : IN  bit;
                 iout   : BUFFER analog_signal := (val => 0.0, kind => 1));

BEGIN
    ASSERT ( imag > 0.0 )
        REPORT "ERROR: charge pump current must be > 0.0."
        SEVERITY failure;

    ASSERT ( imag <= 10.0  )
        REPORT "ERROR: charge pump current must be < 10.0 ma."
        SEVERITY failure;
END charge_pump;

-- body
ARCHITECTURE charge_pump_behavior OF charge_pump IS

  BEGIN
    iout.kind   <= 1;
    iout.val    <= imag AFTER t_delay WHEN (up = '1' AND down = '0') ELSE
                   0.0  AFTER t_delay WHEN (up = '0' AND down = '0') ELSE
                  -imag AFTER t_delay WHEN (up = '0' AND down = '1') ELSE
                   0.0  AFTER t_delay;

END charge_pump_behavior;
```

Figure 19.   Charge Pump Model

The charge pump entity declaration defines two logic input ports, *up* and *down*, and one analog current output, *iout*. The two VHDL generics, *imag* and *t_delay*, customize the output current magnitude and time delay for the charge pump model, respectively.

The architectural body consists of one conditional signal assignment statement for determining the value of the charge pump output current, *iout*. Current is driven after a *time_delay* only when *up* and *down* are not the same logic value. Finally, this current drives the second-order filter described next.

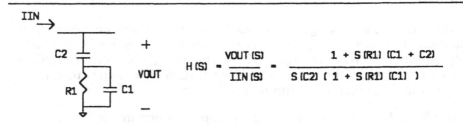

Figure 20.   Second-Order Filter Block Diagram and Transfer Function

4.4.4 Second-Order Filter (F)

The filter shown in Figure 20 is a two-pole one-zero filter. The output voltage (VOUT) is based on both the integral and instantaneous value of the input current (IIN). The output voltage increases for current into the filter and decreases for current out of the filter. The filter has a low frequency zero for instantaneous response and a high frequency roll-off.

Dynamically, for a current input (IIN) the filter integrates the current on the capacitor (C2) and also responds instantaneously to the current passing through the resistor (R1). The output voltage increases for current into the filter and decreases for current out of the filter.

The VHDL model for the second-order filter is shown in Figure 21.

```
USE work.analog_data_structure.all;
USE work.analog_passive_behaviors.all;

-- interface
ENTITY vfo_filter  IS
    GENERIC (    r1: analog_component := ( val => 0.0, kind => res );
                 c1: analog_component := ( val => 0.0, kind => cap );
                 c2: analog_component := ( val => 0.0, kind => cap ) );
    PORT    (    iin : IN      analog_signal;
                 vout: BUFFER analog_signal := (val => 2.0, kind => v));

BEGIN
    ASSERT ( (r1.val > 0.0) AND (c1.val > 0.0) AND (c2.val > 0.0) )
        REPORT "ERROR: R and C's must be > 0.0"
        SEVERITY failure;
END vfo_filter;

-- body
ARCHITECTURE vfo_filter_behavior OF vfo_filter IS
        SIGNAL pvout, p2vout: analog_signal :=
                ( val => 2.0, kind => v );

        SIGNAL piin: analog_signal :=
                ( val => 0.0, kind => i );

    BEGIN
        vout    <= vfof(r1, c1, c2, piin, iin, p2vout, pvout);
        pvout   <= vout  AFTER analog_time_delta;
        piin    <= iin   AFTER analog_time_delta;
        p2vout  <= pvout AFTER analog_time_delta;

END vfo_filter_behavior;
```

Figure 21.   Second-Order Filter Model

The second-order filter model is also derived much like the single pole example described in the previous chapter. In the entity declaration, the current input, *iin*, and voltage output, *vout*, are defined along with the VHDL generics for the filter capacitor and resistor values. In the architectural body, the output voltage is initialized to a reasonable starting value and the finite-difference equation is calculated with the *vfof* function described in Figure 22.

```
--
--
-- function for VFO filter - (2 poles 1 zero)        vout-C2-C1-gnd
--                                                          -R1-gnd
--
    FUNCTION vfof(  r1          : analog_component;
                    c1          : analog_component;
                    c2          : analog_component;
                    pred1_iin   : analog_signal;
                    iin         : analog_signal;
                    pred2_vout  : analog_signal;
                    pred1_vout  : analog_signal )
                              RETURN analog_signal IS

        VARIABLE   num1, num2, num3, num4 :  real;
        VARIABLE   k1  , k2  , k3         :  real;
        VARIABLE   denom                  :  real;
        VARIABLE   vout                   :  analog_signal;

    BEGIN
        k1   := ( r1.val * c1.val * c2.val );
        k2   := ( c2.val );
        k3   := ( r1.val * (c1.val + c2.val) );

        num1 := ( (analog_time_delta_real*analog_time_delta_real +
                    k3*analog_time_delta_real)* iin.val );
        num2 := ( (k3*analog_time_delta_real) * pred1_iin.val );
        num3 := ( (2.0*k1) + (k2*analog_time_delta_real) ) *
                    pred1_vout.val;
        num4 := ( (k1) * pred2_vout.val );

        denom:= ( (k1) + (k2*analog_time_delta_real) );

        vout.val:= ( (num1 - num2 + num3 - num4)/denom );
        vout.kind := v;

        RETURN vout;
    END vfof;
```

Figure 22. Vfof Function for Second-Order Filter

The *vfof* function calculates the next value of voltage based on the previous voltage value **pred1_vout**, its predecessor **pred2_vout**, the current **iin** and its previous value **pred1_iin**. Then, it returns the output voltage value **vout** that drives the VCO.

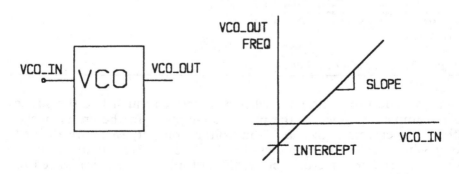

Figure 23. VCO Block Diagram and Transfer Curve

4.4.5 Voltage Controlled Oscillator (VCO)

The VCO shown in Figure 23 has a logic output (VCO_OUT). The frequency of the logic output is a linear function of the voltage input (VCO_IN). The typical oscillator consists of an adjustable current source and a timing capacitor.

Dynamically, the current source charges the internal timing capacitor. As the voltage on the capacitor exceeds a threshold, the logic output changes state and the current direction changes, charging the capacitor in the opposite direction. The current source varies linearly with applied input voltage resulting in a linear frequency -vs-voltage characteristic.

The VHDL model for the VCO is shown in Figure 24.

```
USE work.analog_data_structure.all;
USE work.analog_active_behaviors.all;

-- interface
ENTITY vco  IS
        GENERIC (   slope    : real := 0.100; --ghz/v
                    intercept: real := 0.001; --ghz intercept
                    vhi      : real := 1.0;   --vco cap high volt
                    vlo      : real := 0.0 ); --vco cap low volt
        PORT    (   vco_in  : IN    analog_signal;
                    vco_out : BUFFER bit);

BEGIN
        ASSERT ( (vco_in.val * slope) > (-1.0 * intercept) )
            REPORT "ERROR: negative or zero vco frequency was requested"
            SEVERITY failure;

        ASSERT ( (vco_in.val * slope) < ( 800.0 * intercept) )
            REPORT "ERROR: too high a frequency was requested"
            SEVERITY failure;
END vco;

-- body
ARCHITECTURE vco_behavior OF vco IS
        SIGNAL   v_cap, v_flag : analog_signal :=
                 (val => 0.1E-3, kind => v);

    BEGIN
P1:     PROCESS(v_cap, vco_in)
        BEGIN
            v_cap.val <=   vco_vcap (v_cap, vco_in, slope, intercept,
                                vhi, vlo, vco_out) AFTER analog_time_delta;
            v_flag.val<=   vco_vcapf(v_cap, vco_in, slope, intercept,
                                vhi, vlo, vco_out);
        END PROCESS P1;
B1:     BLOCK( (v_flag.val >= vhi) OR (v_flag.val <= vlo) )
        BEGIN
            vco_out <= GUARDED
                    '1' WHEN ((v_cap.val >= (vhi-vlo)/2.0) ) ELSE
                    '0';

        END BLOCK B1;
END vco_behavior;
```

Figure 24.   Voltage Controlled Oscillator Model

The entity declaration for the model defines the interface. First, the input, *vco_in*, is an analog voltage and the output, *vco_out*, is a logic

type making the VCO an analog-digital behavior. Several VHDL generics customize the VCO. The slope and intercept for the control
transfer curve are set by *slope*, and *intercept*, respectively. Also, the
generics *vhi* and *vlo* set the high and low threshold point on the timing
capacitor internal to the VCO. Assertions check for negative frequency or extremely high frequency during simulation.

The architectural body contains the algorithm for the model. First,
the process P1 is sensitive to changes in voltage on the *vco_in* input
or timing capacitor *v_cap*. The *vco_cap* function returns the value of
voltage on the timing capacitor. The *vco_capf* function returns the
projected value beyond the switching threshold *vhi* or *vlo* when these
levels are exceeded. Block **B1** is sensitive to this projected value and
determines the *vco_out* value based on the exceeded threshold. A signal assignment statement internal to block *B1* compares the timing
capacitor voltage against its midpoint voltage for *v_cap*. The details
of the *vco_vcap* and *vco_vcapf* functions are described next.

```
-- function for VCO - returns the total vco cap voltage
   FUNCTION vco_vcap  ( vcap      : analog_signal; --v
                        vin       : analog_signal; --v
                        slope     : real;          --ghz/v
                        intercept : real;          --ghz
                        vhi       : real;          --v
                        vlo       : real;          --v
                        vco_out   : bit        )
                             RETURN real IS

   VARIABLE   ptime     :  real; --ns
   VARIABLE   v_delta   :  real; --v
   VARIABLE   v_cap_tot :  real;
   VARIABLE   pfreq     :  real; --ghz

   BEGIN
     pfreq   := ( slope * vin.val ) + intercept;  --desired frequency
     ptime   := ( 0.5/pfreq );                    --half period time

     v_delta := ( (vhi-vlo)/ptime ) * analog_time_delta_real; -- + delta v

     IF (vco_out = '1') THEN
         v_delta := -1.0 * v_delta;               --ramp down not up
     END IF;
     v_cap_tot := vcap.val + v_delta;             -- projected vcap
     IF (v_cap_tot > vhi ) THEN
         v_cap_tot := vcap.val-(v_delta - ( 2.0*(vhi - vcap.val) ));
     END IF;
     IF (v_cap_tot < vlo ) THEN
         v_cap_tot := vcap.val-(v_delta - ( 2.0*(vlo - vcap.val) ));
     END IF;
     RETURN v_cap_tot;
   END vco_vcap;
```

Figure 25.   Vco_vcap Timing Capacitor Function for VCO

The *vco_vcap* function shown in Figure 25 calculates the target frequency *pfreq* based on *slope, intercept*, and the current input voltage
vin. It calculates half the period *ptime* and determines the voltage
change *v_delta* for the timing capacitor based on its high and low
levels, *vhi* and *vlo*. Then it adds the delta to the current capacitor

voltage *v_cap_tot* if the output is logic "0", or subtracts the delta if the output is logic "1". If either switching threshold is exceeded, *vhi* or *vlo*, the amount exceeded is subtracted from the threshold level to remove error due to finite step size. Finally, the function returns the voltage value.

```
--
FUNCTION vco_vcapf ( vcap       : analog_signal; --v
                     vin        : analog_signal; --v
                     slope      : real;        --ghz/v
                     intercept  : real;        --ghz
                     vhi        : real;        --v
                     vlo        : real;        --v
                     vco_out    : bit )
                            RETURN real IS

     VARIABLE    ptime      : real; --ns
     VARIABLE    v_delta    : real; --v
     VARIABLE    v_cap_flg  : real;
     VARIABLE    pfreq      : real; --ghz

  BEGIN
     pfreq   := ( slope * vin.val ) + intercept; --desired frequency
     ptime   := ( 0.5/pfreq );                   --half period time

     v_delta := ( (vhi-vlo)/ptime ) * analog_time_delta_real; -- + delta v

     IF (vco_out = '1') THEN
        v_delta := -1.0 * v_delta;               --ramp down not up
     END IF;
     v_cap_flg := vcap.val + v_delta;          -- projected vcap
     IF (v_cap_flg > vhi ) THEN
        v_cap_flg := (vhi + 0.1);
     END IF;
     IF (v_cap_flg < vlo ) THEN
        v_cap_flg := (vlo - 0.1);
     END IF;
     RETURN v_cap_flg;
  END vco_vcapf;
```

Figure 26. Vco_vcapf Timing Capacitor Flag Function for VCO

The *vco_vcapf* function shown in Figure 26 is identical to *vco_cap* except for one aspect. If either switching threshold is exceeded, *vhi* or *vlo*, the amount exceeded is maintained and 0.1 is added. In this way, its value flags an exceeded threshold. It then returns the voltage value used to activate the **B1** block.

4.5 APPLICATION OF PHASED-LOCKED LOOP

Using these models PL loops can be simulated. Changing parameters is easily accomplished by modifying VHDL generics. The following network example illustrates one application.

4.5.1 Network Example

These models form a basis for simulating PL loops in mixed analog-digital chip designs. An example network using these models is shown in Figure 27.

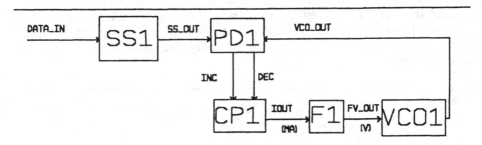

Figure 27. PL Loop Example

The single shot (SS1), phase/frequency detector (PD1), charge pump (CP1), second-order Filter (F1), and voltage controlled oscillator (VCO1) are used. The salient part of the VHDL model for this simulation example is shown in Figure 28.

```
USE work.analog_data_structure.all;

-- interface
ENTITY main_pll IS
END main_pll;

-- body
ARCHITECTURE main_pll_sample OF main_pll IS
     SIGNAL fv_out : analog_signal := (val => 2.0, kind => v);
     SIGNAL iout   : analog_signal := (val => 0.0, kind => i);
     SIGNAL data_in: bit              := '1';
     SIGNAL ss_out, inc, dec, vco_out  : bit := '0';
-- Partial Declarative Region
-- instantiate
     BEGIN
        data_in <= '1' AFTER 40 ns WHEN(data_in='0') ELSE
                   '0' AFTER 10 ns;

        ss1 : single_shot GENERIC MAP ( pulse_width => 25 ns,
                                        t_delay     => 1 ns)
                          PORT MAP    ( data_in, ss_out);

        pd1: phase_freq   PORT MAP    ( ss_out, vco_out, inc, dec );

        cp1: charge_pump GENERIC MAP ( imag => 0.5, t_delay => 1 ns )
                         PORT MAP    ( inc, dec, iout);

        f1:  vfo_filter   GENERIC MAP ( r1 => (val =>  1.0, kind => res),
                                        c1 => (val =>  5.0, kind => cap),
                                        c2 => (val => 24.0, kind => cap) )
                          PORT MAP    ( iout, fv_out);

        vco1: vco         GENERIC MAP ( slope     => 0.010,
                                        intercept => 0.001,
                                        vhi       => 1.0  ,
                                        vlo       => 0.0   )
                          PORT MAP    ( fv_out, vco_out);

END main_pll_sample;
```

Figure 28. PLL Network Entity

The component instantiations correspond directly to Figure 27. The single shot (SS1) has a 1 ns delay and 25 ns pulse width (50% duty cycle at 20 MHZ). The charge pump (CP1) has a 1 ns delay and sources or sinks 0.5 mA. The second-order filter (F1) is comprised of one 1.0 Kohm resistor, and two capacitors of 5.0 pf and 24.0 pf, respectively. Finally, the VCO (VCO1) has a slope of 10 MHZ/volt with a 1 MHZ intercept, and switching threshold limits of 1.0 V and 0.0 V for the timing capacitor.

4.5.2 Simulation Results

The resulting waveforms from the VHDL simulation are shown in Figure 29. Note that these waveforms accurately reflect the theoretical behavior expressed in the PL loop behavioral models.

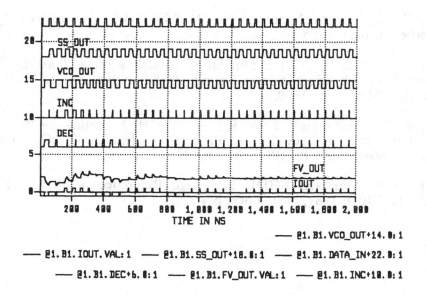

Figure 29.   PLL Example Waveforms

4.5.3 Limitations and Usage

The limitations and usage considerations are the same as those described in the previous chapter. The same cautions regarding time step to relative time constant still apply.

4.6 SUMMARY

This chapter provided detailed modeling applications for the behavioral analog concepts presented in the previous chapter through two common analog-digital loops. Expanding on the models and techniques presented, a mixed analog-digital chip containing these types of loops was simulated. Incorporating this level of detail in what is considered typically a digital simulation model further integrates the analog-digital simulation environment.

REFERENCES

[1] "VHDL User's Manual," No. IR-MD-065-1, Intermetrics Inc., Bethesda, MD, 1985.

[2] B. R. Stanisic, "VHDL Simulation Modeling for a Phase-Locked Loop," VHDL Users' Group Spring 1990 Meeting, pp. 7.01 - 7.13, Boston, MA, Apr. 1990.

[3] B. R. Stanisic and M. W. Brown, "VHDL Modeling for Analog-Digital Hardware Designs," in Proc. IEEE Int. Conf. Computer-Aided Design, pp. 184-187, Nov. 1989.

[4] "IEEE Standard VHDL Language Reference Manual," IEEE Standard 1076-1987, New York, NY.

Chapter 5
Modeling Style Issues
For Synthesis

Steve Carlson

Synopsys, Inc.
Mountain View, California

A new design paradigm, HDL design coupled with synthesis, is quickly emerging as the way to deal with complex ASIC system design. HDLs (hardware description languages) provide designers with the ability to describe a design's architectural and functional characteristics at a more abstract level of representation than a traditional gate level or schematic design methodology allows. Synthesis provides an automated mechanism to generate an optimized gate level representation from an HDL description. The complexity of today's designs mandates the use of higher level, productivity improving methodologies. This chapter examines the technology underlying an HDL-based synthesis methodology and how this new methodology can be applied in practice to an ASIC design.

Although many HDLs exist, VHDL has emerged as the industry standard. The use of VHDL has been adopted as a standard by ANSI, IEEE, NIST, FIPS as well as other European groups and has been required for use on digital designs by the defense department. The examples and the accompanying discussion in this chapter should provide insight into the practical application of VHDL to a new design methodology.

5.1 What is HDL Synthesis?

Most simply stated, HDL synthesis is the process of turning an abstract, technology independent, text description of a design into gates (shown in figure 1). Synthesis is composed of two crucial elements:

Translation: Acts as the automated bridge between two levels of abstraction. In the case here: register transfer level (RTL) and gate level.

Optimization: Technology-specific design transformations to meet goals for the design. In the context here: performance, area, and test.

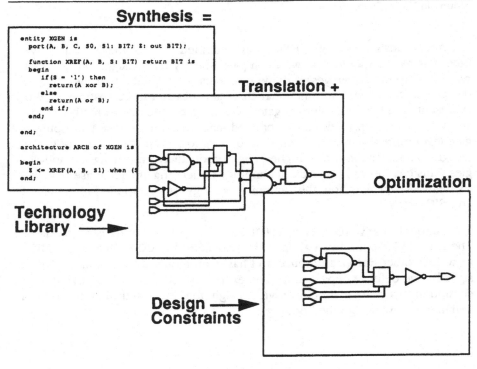

Figure 1. Synthesis Defined

The HDL design methodology presented here encompasses the development of a design description at an RTL level, the validation of that description, a synthesis step where the HDL description is transformed into an optimized gate level representation, and a final verification step. An important new technology enabling this design paradigm is synthesis; as shown in figure 2, synthesis is at the heart of this HDL design methodology.

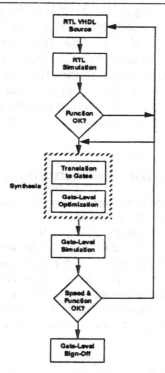

Figure 2. Synthesis Methodology

The key to successful application of automatic synthesis technology is the use of a total design methodology that will yield predictable, accurate, and high quality results. To be successful, a final gate level design needs to meet the functional, area and timing goals of the project, and meet them on schedule. The methodology outlined in this chapter has numerous benefits that will enhance the success of any digital ASIC design team.

Although beyond the scope of this book there are related aspects of synthesis technology that can be applied to the digital design environment. One of these techniques, called Full Circle Synthesis (FCS), embodies a top-down and a bottom-up synthesis process. These two capabilities combined yield a mechanism that has multiple uses. A tool that has FCS capability and can read and write multiple design languages serves as a design gateway for migration between design environments. FCS provides level shifting in the abstraction hierarchy which can be used to translate between languages that have disparate abstraction capabilities (e.g., EDIF and VHDL).

Migration of existing designs into the VHDL design environment is a problem for all new adopters of VHDL, but an FCS capability can provide the automation necessary to ease the burden of the migration task. FCS can be applied to the reprocurement problem faced by the DoD. Another novel use of FCS is the ability to always maintain functionally consistent models between levels of abstraction (technology independent or technology dependent); no matter which level of abstraction the design changes are most easily performed, FCS can generate the corresponding representation in the alternate level of abstraction.

5.2 Applying HDL Synthesis Technology

There are many considerations necessary to understand how, when and why HDL synthesis technology can be applied. The primary considerations actually stem from practical considerations rather than solely from technical innovation. When considering a change in design methodology, organizations typically like to proceed in an evolutionary rather than a revolutionary manner. There are some parallels between the introduction of HDL synthesis technology and other significant advances in CAE technology made in the past. Although HDL synthesis is a revolutionary technology in terms of innovative achievement, it represents only an incremental change to existing design flows.

5.2.1 The Synthesis Continuum

In design automation, synthesis refers to the process of transforming a design from one level of abstraction to another and preferably in a highly efficient manner. As this implies, the overall product design process encompasses many synthesis steps. A synthesis tool is characterized by the levels of abstraction for which it provides an automatic bridge. The HDL Synthesis methodology presented here implies the use of a tool that transforms an HDL description, written at an RTL abstraction level, to produce an optimized gate level netlist of an ASIC design; that is, it is assumed that the result of the HDL Synthesis process will be used as input to the appropriate layout synthesis tool set. For purposes of this chapter, layout is considered a step performed by the ASIC vendor tool set.

Most people agree upon what is meant by a gate level netlist; however, there seems to be a relatively small amount of agreement as to what constitutes an RTL description. This problem with the design community nomenclature is a result of the continuous nature of design description abstraction. Dissimilar to the demarcations between a gate level netlist and layout descriptions, there is overlap in the definitions of higher levels of digital design description.

Figure 3 depicts some of the design steps associated with the total digital design process. We will make the distinction here between an RTL description and a behavioral description. Any description that has had its operations scheduled (implicitly or explicitly) to particular clock cycles is designated as an RTL description. Design descriptions above that level (pre-scheduling) will be designated as behavioral descriptions.

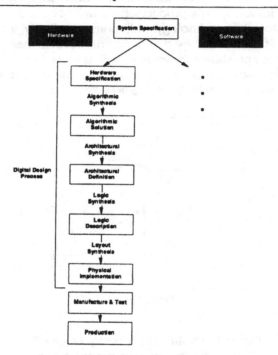

Figure 3. Digital Design Process

In the research community two distinct camps have emerged: logic synthesis and high-level synthesis (also referred to as behavioral synthesis). Logic synthesis spans those design activities below resource allocation down to the technology specific, optimized gate level netlist; that is, it works from the RTL level and below. High-level synthesis covers the spectrum of design problems beginning down at the state machine optimization level and goes up to the highest levels (completely unrestricted) of design. The overlap between high-level synthesis and logic synthesis ensures complete coverage of the design process.

5.2.2 Quality/Productivity Design Automation Acceptance Criteria

Design automation tools are applied to a design problem for two major reasons: to improve design quality and to reduce the design time. Design quality at the gate level is typically measured in terms of design area, speed, and testability. Design time reductions typically are measured as the time to the completion of a project rather than the time for a design iteration (although both are important in a productivity study). The acceptance of a particular design automation technology and, more specifically, a particular tool implementation, is determined by the quality/productivity needs of a design application. The graph shown in figure 4 depicts a two dimensional space of productivity and design quality. The origin of the graph represents the unit productivity and design quality of an engineer with the present configuration of tools.

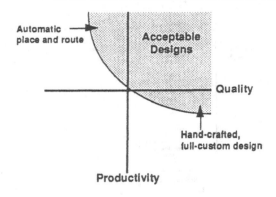

Figure 4. Quality/Productivity Trade-offs

In order for a designer to make a change from this present methodology, he must derive benefit along one or both of the graph axes. Obviously, any tool that represents an improvement in productivity and quality is a clear win and should be adopted. However, there are instances where a designer may be willing to accept a degradation along one axis for improvement along the other axis.

Automatic place and route technology has been widely adopted because of the dramatic increase in designer productivity over manual methods; some estimates are the on the order of ten thousand to one hundred thousand times improvement in this step of the design process. However, it is still acknowledged that, with enough time, a designer working manually could come up with a more efficient solution. Automatic place and route technology thus belongs on the upper right hand edge of the productivity/quality design acceptance curve.

A designer working manually on placement and routing while optimizing transistor sizes can come up with remarkably high performance designs; in some applications, this is important enough to accommodate the added design time to take full advantage of the silicon technology. Instances of this situation are diminishing rapidly. Such activity would be placed at the lower left edge of the productivity/quality design acceptance curve.

As shown in figure 5, logic synthesis fits firmly into the acceptable region of the productivity/quality space. This is due to the relative maturity of logic synthesis technology as compared to high-level synthesis.

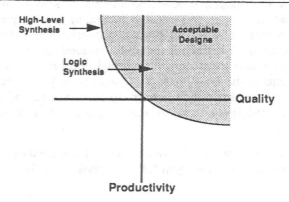

Figure 5. HDL Synthesis Quality/Productivity

Using logic synthesis technology always increases designer productivity. Typical numbers cited for shrinking the gate level design cycle time are in the range from 40 to 60 percent. Logic synthesis can also match or exceed the manual designer in terms of quality of results for most design types. It should be noted that there are classes of highly regular designs that have been extensively studied. In these cases, such as for a multiplier, a designer can usually out perform a general purpose logic synthesis tool. As more such designs are recognized and used, building blocks will exist and be incorporate into the design process.

A problem with much of the research in the area of high-level synthesis is that there has been no strong tie down to the implementation level of detail. The high-level synthesis research community has traditionally been disjoint from the logic synthesis research community. The lack of a link down to the implementation level makes the evaluation of the results of these systems extremely difficult. So, while high-level synthesis promises still more dramatic productivity improvements beyond logic synthesis, current research results lie outside of the acceptable region of the quality/productivity .

It is a given that high-level synthesis is a technology that will be necessary to manage the complexity of the coming design problems. The question is then, how do we get there from the current state of the research? Do we try to extend the research in high-level synthesis down into the logic synthesis domain, or do we extend logic synthesis up into the high-level arena? From a practical standpoint for working engineers, the only viable answer is to work from the bottom-up.

The bottom-up, incremental improvement approach has two principle advantages. The first advantage is it builds on an existing solution in the acceptable quality/productivity space; this ensures general applicability and usefulness in a minimal amount of time for production design use. The second advantage of the bottom-up approach is the ability to feedback information to the high level. The quality measurement of a high level synthesis system must be measured in terms of the final product of the design system, this requires the strong links to silicon that only a bottom-up approach can offer. As design time in logic synthesis is decreased, more understanding of the high level synthesis will occur.

As may be surmised, this chapter will focus on logic synthesis and its use within an HDL environment. The term "HDL synthesis" will be used with this scope in mind.

5.2.3 Practical Considerations

Being able to predict the outcome of any process is fundamental to the most efficient use of that process. This is no less true for HDL synthesis than for a semiconductor processing plant. Just as junction depths must be highly predictable, so must the area and speed results of synthesis. Neither process can tolerate the cost and inefficiencies of sporadic results. For the HDL synthesis task, we can measure "yield" in terms of the circuit quality with respect to a set of design goals. The ingredients of "high yield" synthesis are the input description, the design goals and the optimization algorithms.

The input to HDL synthesis is an RTL functional description of the network to be implemented. As stated above, RTL descriptions are distinct from behavioral level descriptions. The distinction is that a behavioral description has no implied architecture and clocking scheme in its representation, while an RTL level description has a definite, implied architecture. Because of unpredictable resulting circuit quality, it will be some time before true behavioral-level synthesis can be applied in a production environment. However, RTL synthesis has proven itself in the industrial environment on numerous designs.

RTL descriptions produces predictable results during the translation phase of synthesis. But translation is not enough to implement a true synthesis system. Optimization is needed on the translated, technology-specific network to make the area/speed trade-offs necessary to meet the design goals of a particular circuit. Optimization ensures that design quality is not being compromised for designer productivity.

Beyond the pure automation algorithm technology, there are numerous considerations in making an HDL Synthesis system practical for "everyday" design use. The synthesis tool set must be integrated with the rest of the design environment; this means that a wide variety of input and output formats should be supported so that there are easy gateways to existing design tools. The designs produced must be legal designs; that is, vendor specific gate level design rules must be respected. The resultant designs must be testable; hard to test structural implementations are unacceptable because of the negative impact on design schedule and product lifecycle costs of the resulting designs. Because speed is such an important design optimization criteria, a powerful language for specifying design timing constraints is a necessity. To evaluate network adherence to specified timing constraints, a powerful and accurate timing analysis tool is an essential element of a synthesis system. The data for timing analysis needs to come directly from the ASIC vendor to ensure its accuracy and integrity.

There are a number of more general attributes that a Logic Synthesis system should posses. The system should be capable of handling large designs (million gate ASICs are just around the corner). The tool should have broad applicability across design technologies and be capable of making the most efficient use of all of the cells in a particular library. All types of design problems should be addressed, from digital signal processing to CPU design. Lastly, quality of results is mandatory for the tool to be adopted in mainstream design; results must be on par with an expert level designer.

So where does all of this lead? VHDL is a rich and expressive language with features geared for many aspects of the design process. To enable a designer to specify efficient designs, one must understand how to most efficiently communicate design intent to a synthesis system. The knowledge and process involved in achieving this will be called a synthesis policy.

5.3 An HDL Synthesis Policy

A Synthesis Policy encompasses all of the elements necessary to ensure successful HDL synthesis. The purpose of the policy is to define a design methodology that enables you to predictably produce high quality silicon. The tremendous productivity gains afforded by an HDL synthesis methodology are lost if a high quality chip (in terms of both speed and area) is not the end result of the synthesis process. This issue is language-independent and composed of:

1. The design methodology,
2. Style templates for design description styles, and
3. The HDL vocabulary (language constructs) used to support the style and methodology.

Each is covered in further detail below.

5.3.1 Design Methodology

The design methodology to be detailed here guides the flow from an RTL description down to an optimized gate level description. The reader is referred back to figure 2 for a diagram of this flow.

The logic synthesis process begins with a validated RTL design description that has been partitioned down to the implementation primitives of ALU's, RAM, ROM, data paths, and control function. It is assumed that the design's operations have been fully partitioned into clock cycles at this point. Included with the design, expressed in an HDL, is the functional test stimuli and results used to validate the design.

In the synthesis phase of the design cycle, the design is translated to a gate level representation, and optimized with respect to a particular set of design goals and target technology library. The results of the synthesis process is a technology specific, optimized gate level representation. At this point, a verification step takes place.

Verification differs from validation. Validation is a proof by establishing a sufficient body of evidence, while verification is a proof by comparison between a validated description and a test description. The synthesized design is verified against the RTL description using the functional tests.

This validation must be done with the performance characteristics of the target silicon technology taken into consideration to verify that both the desired performance characteristics and functionality have been achieved. If the specifications have not been met, then further optimization may be required, or possibly modifications to the RTL source must be made to try a new architecture that is more amenable to the design goals. If the gate level description does meet the design objectives, then the designer proceeds through the further steps necessary to complete the design.

5.3.2 Design Style

The second component to the synthesis design policy is the style in which the HDL description is written. The style used is a powerful way in which one controls the synthesis process. Similar to writing high-level software, different HDL descriptions of the same functionality can yield radically different results. The power to control the synthesis process is the power to express the design intent. The resultant design quality comes from both the description style used by the engineer and the power of the synthesis tool.

Later in this chapter, examples of typical hardware modules found in ASIC designs are introduced. These examples can be used as style templates for designing similar hardware. There are no strict style guidelines that need to be followed; rather, the examples represent description styles that have been found to be effective starting points for the synthesis process. The intent is to show, by way of example, proven styles of description for representative hardware types.

Areas in which the designer can exert the most influence using style as a control mechanism are in describing sequential timing/clocking schemes, utilizing design hierarchy, or taking advantage of predefined blocks.

Sequential timing/clocking schemes used in a design can drive the synthesis process to produce a more parallel or sequential design. From a high level, ASIC designs are typically thought of in terms of RAM, ROM, control, and data path elements. In turn, control and data path elements are composed of units such as ALUs, register files, state machines, and random logic. This level of decomposition provides the right granularity of control steering the synthesis process towards desired design goals.

Hierarchy is an important way that humans deal with complexity. It allows a complex problem to be partitioned into smaller, more manageable, problems. Hierarchy in digital design has importance beyond complexity management. It can impact the physical implementation of the design, the overall design quality, and has implications in design re-use. One of the great benefits of HDL-based synthesis is that the effects of creating and merging levels of hierarchy can be studied at relatively little cost. With a basic understanding of the ways in which hierarchy is used by the CAE tools downstream from synthesis (e.g., floorplanning and layout), one can use the synthesis process to arrive at a hierarchy that takes best advantage of those features.

Most ASIC design libraries have many large building blocks (multipliers, microcontrollers, etc.). These blocks are typically highly optimized for the particular technology being used, and provide the library user with a resource of high quality "parts" that might help in his design task. The synthesis tool, like the designer, needs to incorporate these previous designs in order to improve the productivity and reliability of the overall design process.

The most important point about style is that it provides a controlling mechanism to meet design objectives.

5.3.3 Supported Language Constructs

The third component of the HDL Synthesis Policy is the expressiveness and clarity of the language constructs used to support the level of description required for RTL synthesis. This is the designer's vocabulary for design description, the words that will form the design description style. An important distinction in the discussion of language support is the difference between the level of description at which a particular construct might be categorized, and the level of description used to describe the design. Constructs that are more typically behavioral (e.g., the "+" operator) can be fully supported within the context of an RTL description and will yield high quality circuits.

A problem that exists for synthesis tools is that most HDLs have been created for other purposes (such as simulation) and not specifically for synthesis. This is a problem only for synthesis tools and is actually a benefit for designers. The designer needs the capability to easily, in terms of both descriptive power and simulation turnaround time, validate each design description. Thus, the designer needs the same language to develop the test environment as well as the description of the hardware to be implemented. Synthesis tools must be able to work around the simulation and validation specific constructs of the language, and concentrate their efforts on those portions of the design description that relate to circuit function.

The exact set of constructs supported for logic synthesis is not yet standardized. This is unfortunate for the designer, as he must choose a vendor before commencing the design process. The reason that it is not yet practical to standardize down to a very detailed level is because of the disparity between the capabilities and maturity of the various development groups. Therefore, to assist the designer, listed below are the general VHDL features that should be avoided when composing a "synthesizable" RTL model.

Those constructs that relate to file input and output during simulation are generally considered to be part of the "unsynthesizable" category of VHDL constructs. The typical use for file I/O is for simulation input stimulus and output response. Therefore, synthesis models should not contain the following constructs: **File** types, **File Declaration**, and **Package TEXTIO**.

The data types supported for synthesis are a direct reflection on the level of abstraction supported by a particular synthesis tool. Pay particular attention to the use of VHDL primitive types: **Physical**, and **Floating Point**. The use of these data types often requires interpretation by the synthesis tool that the designer may not consider desirable.

Assertion statements in VHDL are used primarily as an error checking mechanism for model behavior and do not effect the simulation behavior. As such, constructs such as the VHDL **Assertion Statement** and **Concurrent Assertion Statement** can generally be excluded from the synthesis process.

Other VHDL constructs to avoid for models aimed at a synthesis system are **Allocators** (which are used for dynamic memory allocation), **Access** types and **Incomplete** types (i.e., self or cross referencing types). These constructs provide capabilities and semantics not easily represented in hardware.

VHDL contains two classes of attributes: predefined and user-defined. Many of the predefined attributes are aimed at dealing specifically with simulation issues, and as such should be avoided or contained in their usage. For example: **DELAYED(T), QUIET(T), TRANSACTION, ACTIVE LAST_EVENT**, and **LAST_ACTIVE**.

User-defined attributes can be used in a multitude of ways, most of which are tool dependent. In terms of synthesis tools, attributes are usually used to add information about the design. Typically design goals (e.g., area, clock speed) and constraints (e.g., setup, hold, and maximum fanout) are coded in terms of user-defined attributes.

Unfortunately for the designer, there has been no standardization of the set of attributes or attribute names for synthesis. The justification for using user-defined attributes to specify design goal and constraint information is that all of the design information is contained in one file. The reason against this approach is that putting such information in a technology independent RTL description actually binds it to a particular tool and technology.

Another method of information embedding that is commonly used by synthesis systems is the use of interpreted comments. Such comments, often referred to as synthetic comments, are typically used as directives to the front-end parser. Functions include: ignoring sections of code that are intended only for simulation or model test and various efficiency or error reporting verbosity control.

Now that there is a common basis from which to discuss logic synthesis and RTL descriptions, the interpretation of VHDL constructs during the synthesis process can be explored.

5.4 Synthesis of Register Transfer Level Constructs

This section is intended to take some of the mystery out of the VHDL synthesis process. It is organized as a collection of several small VHDL examples. These examples are constructed to illustrate how logic gates are synthesized from VHDL code.

Each example introduced below consists of a figure and a text description. The figure contains a VHDL source code fragment which contains only one or two VHDL constructs and a schematic showing the synthesized logic having been mapped to a "typical" ASIC library. This is followed by a short paragraph describing the VHDL code and how the code is synthesized.

The first example illustrates the synthesis of VHDL's basic concurrent construct. The entity section defines the name of a VHDL design. The port declarations declare a signal interface to the design. The architecture section declares the function of the design which, in this case, consists of a simple logic equation in a signal assignment

```
entity VHDL is                          ┌─────────────────────────────┐
    port(                               │ Entity and port declarations │
        A, B, C : in  BIT;              │ define interface             │
        Z      : out BIT                └─────────────────────────────┘
    );
end VHDL;

architecture VHDL_1 of VHDL is          ┌─────────────────────────────┐
begin                                   │ Architecture                 │
    Z <= (A and B) or C;                │ declaration defines interface │
end VHDL_1;                             ├─────────────────────────────┤
                                        │ Logic Expression             │
                                        └─────────────────────────────┘
```

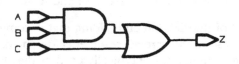

Figure 6. Concurrent VHDL

The next example illustrates the use of VHDL's vectored data types. The synthesis tool will create multiple-bit busses for signals and variables whose types must be represented by more than one bit. This example also illustrates how VHDL processes may contain sequential statements that cause logic to be cascaded together (note the cascaded and-or structure).

```
entity VHDL is
     port(
          A, B, C : in  BIT_VECTOR ( 1 to 5 );  ───  Ports are 5 bits wide
          Z     : out BIT_VECTOR ( 1 to 5 )
     );
end VHDL;

architecture VHDL_1 of VHDL is                    Process groups
begin                                             sequential statements
     process
          variable TEMP : BIT_VECTOR ( 1 to 5 ); ─  Logic Expression
     begin
          TEMP := A and B;
          Z   <= TEMP or C;             ─────  Sequential Statements
     end process;
end VHDL_1;
```

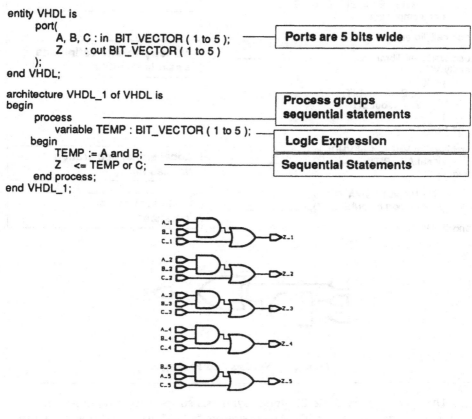

Figure 7. Vector Types

Many times a designer may have an existing design which needs to be re-implemented. Structural-level VHDL may be synthesized even when referencing project package stored in a separate file. This style of VHDL may be used in conjunction with other tools that create VHDL netlist output. This example also illustrates the use of VHDL packages. Packages can be used to group together modeling abstraction layer constructs such as types and conversion functions.

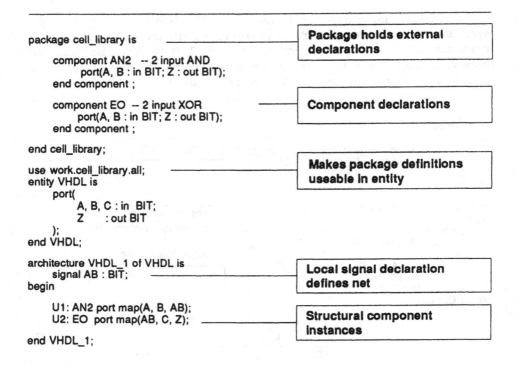

```
package cell_library is                          Package holds external
                                                 declarations

    component AN2   -- 2 input AND
        port(A, B : in BIT; Z : out BIT);
    end component ;

    component EO  -- 2 input XOR               Component declarations
        port(A, B : in BIT; Z : out BIT);
    end component ;

end cell_library;

use work.cell_library.all;                       Makes package definitions
entity VHDL is                                   useable in entity
    port(
        A, B, C : in  BIT;
        Z      : out BIT
    );
end VHDL;

architecture VHDL_1 of VHDL is                   Local signal declaration
    signal AB : BIT;                             defines net
begin

    U1: AN2 port map(A, B, AB);                  Structural component
    U2: EO  port map(AB, C, Z);                  instances

end VHDL_1;
```

Figure 8. Netlists and Packages

The following example illustrates synthesis support for a higher-level arithmetic operator. The arithmetic operators "+", "-" and "abs" operators can normally be supported in their full generality. Also, "*", "/", "**", "mod" and "rem" are usually supported, at least for constant expressions and power of 2 operands.

```
entity VHDL is
     port(
          A, B : in  INTEGER range 0 to 15;
          C    : out INTEGER range 0 to 15
     );
end VHDL;
```

Integer range defines number of bits in a port

```
architecture VHDL_1 of VHDL is
begin
     C <= A + B;
end VHDL_1;
```

Adding two integer values

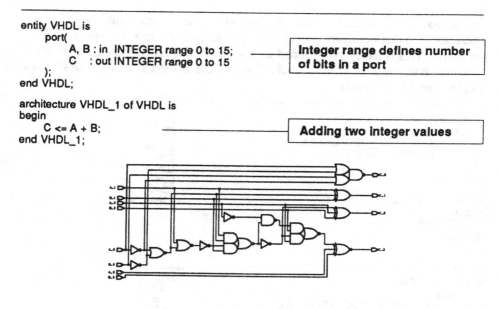

Figure 9. Arithmetic Operators

Comparison operators can also be supported. Note the "<" operator in VHDL can imply a large amount of logic. The comparison operators, "<", ">", "<=", ">=", "/=" and "=" can be supported in their full generality. An additional note to point out is that VHDL does not define the numeric interpretation of **bit_vector** type. Therefore, the "<" operator is composed of different logic depending on the interpretation. This must be clarified in the code or a package if correct designs are to be simulated and synthesized.

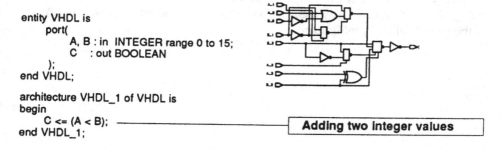

```
entity VHDL is
     port(
          A, B : in  INTEGER range 0 to 15;
          C    : out BOOLEAN
     );
end VHDL;
```

```
architecture VHDL_1 of VHDL is
begin
     C <= (A < B);
end VHDL_1;
```

Adding two integer values

Figure 10. Relational Operators

Indexing into an arrayed object is not just a simple selection. Notice that there are two distinct cases considered. In the first example, the index is a constant. This is easily synthesized as a wire. In the second example the index may take on any value. A multiplexor of the correct size is synthesized to implement this construct.

```
entity VHDL is
    port(
        A     : in BIT_VECTOR( 0 to 7 );
        OUTPUT : out BIT
    );
end VHDL;

architecture VHDL_1 of VHDL is
begin
    OUTPUT <= A( 5 );
end VHDL_1;
```

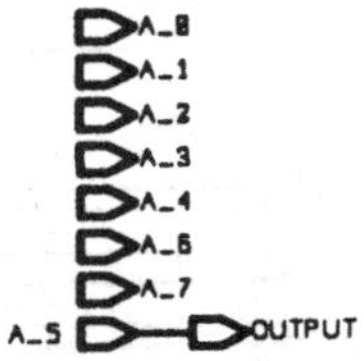

```
entity VHDL_VAR is
    port(
        A     : in BIT_VECTOR( 0 to 7 );
        INDEX  : in INTEGER range 0 to 7;
        OUTPUT : out BIT
    );
end VHDL;

architecture VHDL_1 of VHDL_VAR is
begin
    OUTPUT <= A( INDEX );
end VHDL_1;
```

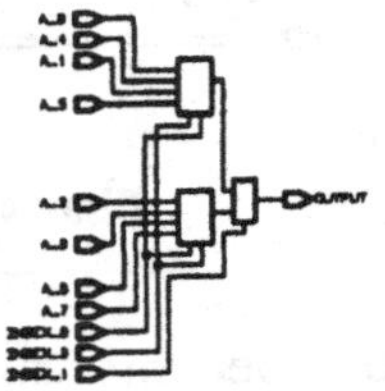

Select from array based on port input

Figure 11. Vector Indices

The if-then-else statement is used in VHDL to conditionally execute sequential statements. For the purposes of synthesis, this statement implies a multiplexing of signals. Any signal or variable that is assigned to within the "if-then-else" will be multiplexed.

```
entity VHDL is
      port(
            A, B  : in  BIT;
            USE_B : in  BIT;
            Z     : out BIT
      );
end VHDL;

architecture VHDL_1 of VHDL is
begin
      process begin
            if (USE_B = '1') then
                  Z <= B;
            else
                  Z <= A;                          ──────── If-then-else statement
            end if;
      end process;
end VHDL_1;
```

Figure 12. If-Then-Else

Like the "if-then-else" statement, a "case" statement implies a multiplexer. The following example illustrates the use of a user-defined type (called ENUM). Note that this example shows the use of a special synthesis attribute which optionally specifies how the user-defined type should be encoded in the synthesized hardware.

```
package types is
      type ENUM is (USE_A, USE_B, USE_C, USE_D);
      attribute ENUM_ENCODING of ENUM : ──────  User-defined type (enumerated
            type is "00 01 10 11";                value) with optionally supplied
end types;                                        user encoding for the type

use work.types.all;
entity VHDL is
      port(
            A, B, C, D : in  BIT;
            CHOICE     : in  ENUM;
            Z          : out BIT
      );
end VHDL;

architecture VHDL_1 of VHDL is
begin
      process (CHOICE) begin
            case CHOICE is ──────────────────  Case statement
                  when USE_A =>  Z <= A;
                  when USE_B =>  Z <= B;
                  when USE_C =>  Z <= C;
                  when USE_D =>  Z <= D;
            end case;
      end process;
end VHDL_1;
```

Figure 13. Case Statement

The VHDL "for" statement may be used like a macro to iterate sections of logic. In this example, the "for" loop is used to create an 8-input exclusive or. The for loop is generally useful for creating custom adders, comparators, and other regular structures. For most synthesis systems, the loop bounds must be statically computable. If not, the loop would have to be implemented as a sequential state machine, leading to expensive analysis and undesirable results.

```
entity VHDL is
    port(
        WORD    : in  BIT_VECTOR ( 0 to 7 );
        PARITY  : out BIT
    );
end VHDL;

architecture VHDL_1 of VHDL is
begin
    process
        variable RESULT : bit;
    begin
        RESULT := '0';
        for I in 0 to 7 loop  ─────────────── for loop to xor all bits
            RESULT := RESULT xor WORD(I);       In "WORD"
        end loop;
        PARITY <= RESULT;  ──────────────────  Assign result to output
    end process;
end VHDL_1;
```

Figure 14. For Statement

This example illustrates several VHDL constructs, including subtypes, slices, and functions. The synthesizer will create hardware for a function whenever that function is called. Functions can be called several times and can be nested. Recursion, bounded by a constant, can also be supported by many synthesis systems.

```
package ops is
    subtype WORD is BIT_VECTOR (1 to 16);  ─────  Sub-type declaration
    function asr(INPUT : WORD) return WORD;  ────  Function declaration
end ops;

package body ops is   -- Arithmetic shift right function
    function asr(INPUT : WORD) return WORD is  ──  Function body
        variable RESULT : WORD;
    begin
        RESULT(1)      := INPUT(1);
        RESULT(2 to 16) := INPUT(1 to 15);  ──────  Use of array "slices"
        return RESULT;
    end;
end ops;
```

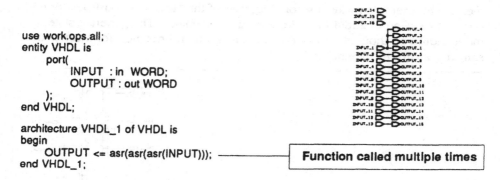

```
use work.ops.all;
entity VHDL is
    port(
        INPUT  : in  WORD;
        OUTPUT : out WORD
    );
end VHDL;

architecture VHDL_1 of VHDL is
begin
    OUTPUT <= asr(asr(asr(INPUT)));  ──────┤  Function called multiple times
end VHDL_1;
```

Figure 15. Subtypes, Slices, and Functions

The use of a wait statement implies to the synthesis tool that the value assigned to signals in a process must be stored. Registers are inserted with the wait condition forming the enabling clocking function. In the example shown, the value of the signal TOGGLE is preserved between clock cycles. As will be shown later, this construct is one of the most important for controlling the interpretation of the VHDL code.

```
entity VHDL is
    port(
        ENABLE : in  BIT;
        CLOCK  : in  BIT;
        TOGGLE : buffer BIT
    );
end VHDL;

architecture VHDL_1 of VHDL is
begin
    process begin
        wait until not CLOCK'stable and CLOCK = '1';  ──────┐
        if (ENABLE = '1') then
            TOGGLE <= not TOGGLE;
        end if;                           ┌──────────────────────────────────┐
    end process;                          │  wait statement infers a flip-flop │
end VHDL_1;                               └──────────────────────────────────┘
```

Figure 16. Wait Statement

The concurrent generate statement is used to iteratively replicate a number of concurrent statements. The loop index is used to customize each statement. The example makes use of a number of the language constructs previously introduced in addition to the generate statement. The generate statement should be thought of as a convenient shorthand notation for a series of similar concurrent statements.

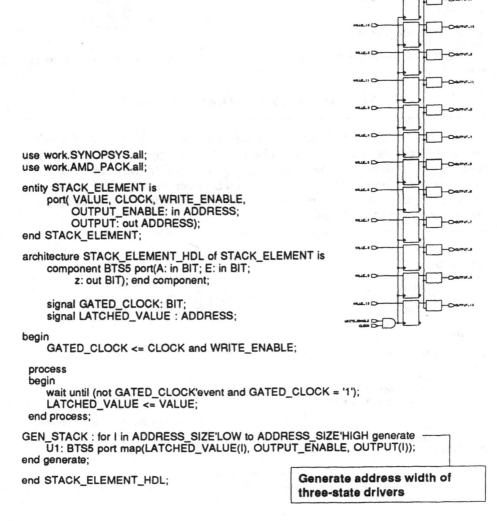

```
use work.SYNOPSYS.all;
use work.AMD_PACK.all;

entity STACK_ELEMENT is
    port( VALUE, CLOCK, WRITE_ENABLE,
        OUTPUT_ENABLE: in ADDRESS;
        OUTPUT: out ADDRESS);
end STACK_ELEMENT;

architecture STACK_ELEMENT_HDL of STACK_ELEMENT is
    component BTS5 port(A: in BIT; E: in BIT;
        z: out BIT); end component;

    signal GATED_CLOCK: BIT;
    signal LATCHED_VALUE : ADDRESS;

begin
    GATED_CLOCK <= CLOCK and WRITE_ENABLE;

    process
    begin
        wait until (not GATED_CLOCK'event and GATED_CLOCK = '1');
        LATCHED_VALUE <= VALUE;
    end process;

GEN_STACK : for I in ADDRESS_SIZE'LOW to ADDRESS_SIZE'HIGH generate
    U1: BTS5 port map(LATCHED_VALUE(I), OUTPUT_ENABLE, OUTPUT(I));
end generate;

end STACK_ELEMENT_HDL;
```

> **Generate address width of three-state drivers**

Figure 17. Generate Statement

A synthesis tool can be controlled to generate different circuit realizations of the same logical functions by specifying different goals for optimization. In the VHDL_SMALL architecture, a MAX_ AREA attribute causes the design to be optimized with a constrained area. In the VHDL_FAST architecture, a MAX_ DELAY attribute causes the design to be optimized with a constrained delay. Notice that both designs perform the same logical function. They simply have different performance and area characteristics. This begins to introduce one of the key features of a logic synthesis system — the ability to quickly optimize the resultant design.

```
entity VHDL is
    port(
        A, B : in  BIT_VECTOR (1 to 7);
        EQL  : out BOOLEAN
    );
end VHDL;

architecture VHDL_SMALL of VHDL is
    attribute MAX_AREA of VHDL_SMALL : entity is 0; ───────
begin
    EQL <=  (A = B);
end VHDL_SMALL;
```

Attribute controlling optimization: area constraint

```
architecture VHDL_FAST of VHDL is
    attribute MAX_DELAY of EQL : signal is 0; ──────
begin
    EQL <=  (A = B);
end VHDL_FAST_1;
```

Attribute controlling optimization: speed constraint

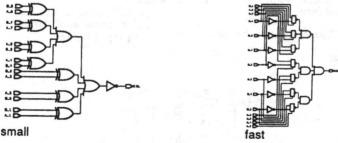

small   fast

Figure 18. Controlling Optimization

The final example brings together many of the constructs discussed earlier. Note that VHDL constructs may be mixed arbitrarily to create a complete design. Note that the combined synthesized logic has been optimized at the gate level. This allows for a much more efficient design than would be produced by simply cascading together translated constructs.

```
package types is
        type ENUM is (STOP, SLOW, MEDIUM, FAST);  ──┤ ENUM declaration
end types;

use work.types.all;
entity VHDL is
    port(
            ACCELERATOR    : in BIT;
            BRAKE          : in BIT;
            CLOCK          : in BIT;
            SPEED          : buffer ENUM  ──────── Output is current state
    );
end VHDL;

architecture VHDL_1 of VHDL is
begin                                              Implies state flip-flops
    process begin
            wait until not CLOCK'stable and CLOCK = '1';
            if (ACCELERATOR = '1') then
                case SPEED is
                    when STOP =>      SPEED <= SLOW;
                    when SLOW =>      SPEED <= MEDIUM;
                    when MEDIUM =>  SPEED <= FAST;
                    when FAST =>      SPEED <= FAST;
                end case;
            elsif (BRAKE = '1') then
                case SPEED is
                    when STOP =>      SPEED <= STOP;
                    when SLOW =>      SPEED <= STOP;
                    when MEDIUM =>  SPEED <= SLOW;
                    when FAST =>      SPEED <= MEDIUM;
                end case;
            else
                    -- Speed does not change          Next state assignment based
            end if;                                   on input value
    end process;
end VHDL_1;
```

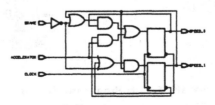

Figure 19. Finite State Machine

This section has introduced the synthesis of specific VHDL constructs into gates. But important global design issues must be thought of when putting these simple language elements together. These global issues, termed synthesis style issues, are covered next.

5.5 Synthesis Style Issues In VHDL

Modeling style provides the key controlling mechanism to meet design objectives. The power to control the synthesis process is the power to express the design intent. The design quality resultant from the HDL synthesis process comes from both the description style used by the engineer and the power of the synthesis tool. The designer always maintains final control over the outcome of the synthesis process through the style of coding used. It is therefore important for the designer to understand how various modeling techniques effect the overall result. The intent of this section is to solidify the ideas shown in the previous section and to put forth the higher level issues in composing an overall HDL description of a complete design.

The primary subsections here relate to the creation of a technology independent modeling paradigm, how system timing issues can be handled, and finally a couple of small design examples to illustrate the synthesis concepts in use.

5.5.1 Process Independent Modeling Paradigm

One of the key benefits of an HDL based methodology is the opportunity to design in a technology independent manner. This means a deferral of the actual silicon technology choice, and an inherently more retargetable design description. Because most new large system designs contain a significant portion of previously designed logic, having "portable" designs is extremely valuable. Introduced here are some of the considerations in writing process independent designs.

The model creation process can be viewed as the interconnection of communicating processes. Processes are connected via signals. Groups of processes can be placed within an embedded entity-architecture pair. Entities are used to create design hierarchy. A diagrammatic view of this paradigm is shown below in figure 20.

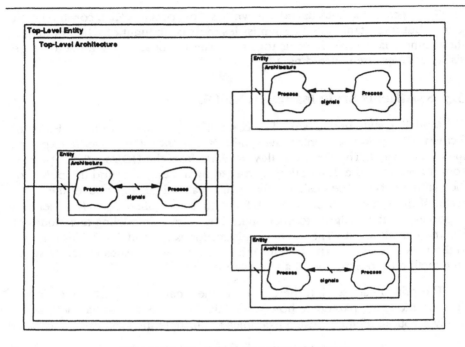

Figure 20. Modeling Paradigm Diagram

Hierarchy is an important method for humans to deal with complexity. It allows a complex problem to be partitioned into smaller, more manageable, problems. Hierarchy in digital design has importance beyond complexity management; it can impact the physical implementation of the design, the overall design quality, and has implications in design re-use.

An important consideration in the preparation of technology independent HDL models is the use of electrically based architectural decisions. Probably the most common example of this type of choice is a bus structure. Three-state buses work great within the MOS technology families because of the utility of pass transistor networks. However, in ECL, multiplexer structures are the implementation of choice. I/O pads are another portion of the design where the decisions made are primarily concerned with electrical rather than functional aspects of the network. When creating "portable" HDL models, it is important to isolate those parts of the design where technology-dependent information affects the functional description. Design hierarchy is an excellent way to achieve the isolation. Recognizing and isolating the portions of the design is an important step in technology- independent HDL model composition.

Clock rate and signal delay are highly process dependent system performance parameters. Placing clock rate and signal delay timing information into technology independent models is self defeating. Embedding such information in the models, even if it is considered a design goal or constraint (rather than actual performance), ties the model to a particular generation of technology. While performance goals are an important ingredient to quality of results in the synthesis process, they are better represented in a contained context (such as VHDL package constants or separate constraint files) aimed at a particular process library. The inclusion of "accurate" timing in high level models also poses a problem because the structure of the HDL code may not bear any resemblance to the physical path structures in the gate level implementation.

An alternative to explicit inclusion of technology dependent timing in high level models is the use of a metric-free timing representation more commensurate with the level of abstraction being used for the representation of function. For synchronous designs, at an RTL level, the natural abstraction is to that of a system clock used for global synchronization. That is, the use of an event oriented synchronization rather than a time unit specification is more appropriate in the RTL domain. The abstraction process allows the deferral of the binding of physical units. The hazard of this abstraction is the isolation from physical reality. The use of synthesis directly compensates for the shortcomings of abstraction by allowing a fast path down into the targeted technologies. The technology specific representations provide the model for accurate analysis.

Asynchronous designs, by their very nature, require more specification than their synchronous counterparts. While synchronous designs can rely on a global event oriented synchronization process, asynchronous designs rely upon specific timing relationships for proper network function. This means that in order to create an accurate portrayal of circuit function all such relative relationships must be specified. The related issues are highlighted by studying storage element descriptions. The issues related to asynchronous design are deferred to a subsequent section.

5.5.2 Synchronous Operation Through Implicit Storage Elements

RTL design has typically been thought of as a paradigm where all storage elements in the network are explicitly defined. A slight push upward in the abstraction ladder allows the use of implicitly defined storage elements in the design description. Looking back at the finite state machine of figure 19, there is no explicit definition of a register. Instead, a port signal is assigned to inside of a global synchronization (wait for clock edge). This assignment encapsulated in a synchronization construct is an explicit method of implicitly identifying a register. The usefulness is that the register inference is accomplished without explicitly using a component with specific ports. A more technology independent description results.

There are also other methods to perform implicit storage element definition in an HDL description. The most straightforward method to describe is the incompletely specified conditional assignment. In this method, a signal is assigned a new value only under the true evaluation of some conditional expression; otherwise, the signal is not assigned to at all. In the latter case where the signal is not assigned to, it should maintain its old value. This implies the creation of state. As an example, the VHDL code fragment in figure 21 and the corresponding schematic show how an incompletely specified conditional assignment implicitly specifies the inclusion of a latch.

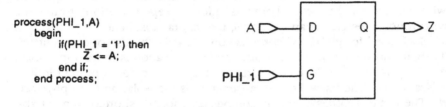

```
process(PHI_1,A)
    begin
        if(PHI_1 = '1') then
            Z <= A;
        end if;
    end process;
```

Figure 21. VHDL Code and Generated Schematic

The capability to implicitly define the storage elements of your design removes you from one level of detail and ensures technology independent storage element description. The hazard is that unnecessary state will be generated because of careless coding practices. Anytime a functional description contains a incompletely specified conditional assignment, state will be generated.

As an illustration of the utility of the implicit state definition, a simple two-phase clock scheme latch design is shown below in figure 22. The clock signals PHI_1 and PHI_2 are externally generated and assumed to be non-overlapping signals. The machine performs a logic function on the incoming signal A, and feeds the output, Z, with the PHI_2 clocked latch. Although a somewhat trivial example, this illustrates a template for design of a two-phase system.

```
entity TWO_PHASE is
    port(PHI_1, PHI_2, A: in BIT;   Z: buffer BIT);
end TWO_PHASE;

architecture DUAL_PHASE of TWO_PHASE is
    signal TEMP: BIT;
begin
    process(PHI_1,A,Z) begin
        if(PHI_1 = '1') then
            TEMP <= A and Z;
        end if;
    end process;

    process(PHI_2,TEMP) begin
        if(PHI_2 = '1') then
            Z <= not TEMP;
        end if;
    end process;
end;
```

Figure 22. VHDL Code and Generated Schematic

The PHI_1 process contains all of its logic in this example. It is often convenient to set up input and output vectors to the processes to separate the
logic and storage functions of the design. That is not a mandatory part of the
methodology, but it can often lead to a "cleaner" model.

5.5.3 Partially Asynchronous Operation

Often in the design of synchronous systems, partial asynchronous operations are a convenient or required part of the specification. Typical examples
are a system wide reset on power-up or partial system state reset on error or
failure. This section introduces a straightforward style for the description of
partially asynchronous networks.

It is simple to turn a Mealy state machine into a partially asynchronous
machine. In this case we wish to add a new port, RESET, to put the state machine back into state S0 when the signal is high. The new process description
and complete schematic are shown here in figure 23.

```
— Process to hold synchronous elements (flip-flops)
SYNCH: process(clock,reset)
    begin  — asynchronous condition to go back to S0
        if (RESET = '1') then
            CURRENT_STATE <= S0;  — else normal synchronous operation
        elsif (CLOCK'event and CLOCK = '1') then
            CURRENT_STATE <= NEXT_STATE;
        end if;
    end process;
```

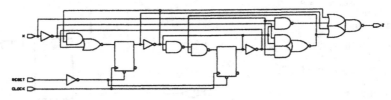

Figure 23. VHDL Code and Generated Schematic

The use of partially asynchronous operations is essential in most digital designs, which makes such descriptive capability a mandatory part of an HDL-based synthesis policy. This capability, however, can be a source of problems if not used descriminately. For instance, it now becomes possible to describe networks with highly complex race conditions; the asynchronous part of the sequential assignments can be chained via communicating signals between two processes resulting in interdependent asynchronous operation. It is much safer to handle the communication between two processes via synchronized signals.

5.5.4 Asynchronous Operation

Most high performance systems today contain at least one subsystem that operates asynchronously. Although complicated to deal with, the overall increased speed of the implemented hardware warrants their use. The basic consideration in HDL synthesis based asynchronous design is that the execution of operations is controlled via completion and initiation signals. The machines are designed such that the completion of one operation initiates the execution of the next consecutive operation (which is determined by the present state and/or input values). This concept is illustrated more fully in the complete example of the final section.

5.6 A Complete Example

The operation characteristics of an 8-bit UART make it an ideal example for showing how to approach an asynchronous design problem and provides a complete design example of the synthesis process. The UART design example here is based on worked completed at Virginia Tech. by Dr. James Armstrong. The two main functions of a UART are: 1) take a parallel data word in from the computer and write it out serially and, 2) read in serial external data and convert it to a parallel word for the computer. These functions are described as follows:

The serial-in parallel-out operation is initiated by a negative transition on a serial data input that is held low for over half a clock period. The half period time mark is then used to clock in a data word (8 bits in this example). When an entire word is received, the completion signal NINTI is set low until a data word has been read.

The parallel-in serial-out operation is initiated by a positive transition of a LOAD signal. Completion of the serial-out operation is signalled via a low state of NINTO. A common parallel word interface port, DATA, is used for external input of parallel-in data and parallel-out data.

The design problem is most easily understood when approached hierarchically. The UART is broken into four major functional subunits shown in figure 24. Each of these units is represented in its own code section as described next.

PAR_IN_SER_OUT	- parallel-in serial-out operation
SER_IN_PAR_OUT	- serial-in parallel-out operation
CLOCK_GEN	- clock generator
BTS4_BANK	- bidirectional I/O control

Figure 24 Subunits of the UART

A project package lays the foundational work for the network description of the UART subunits and is shown is figure 25. To make the model more general the word length (WORD_LENGTH) is set by using a VHDL constant in the package. Although the DATA port of the design is bidirectional, the type BIT can still be used as the base data type for the system. To have multiple drivers (sources) on a signal, a bus resolution function (BRF) called WIRED_OR is defined for simulation purposes. The BRF can be ignored by the synthesis tool because a simple wired connection is all that is actually needed. Although it would be nice to use the BRF to define a port driver of a synthesized component, this is beyond the scope of the current implementation. It is also convenient to put all of the component statements into the package so they do not have to be repeated throughout the code.

```
package TSL_Package is

    constant WORD_LENGTH : integer := 8;
    subtype WORD is BIT_VECTOR((WORD_LENGTH-1) downto 0);
    type WORD_VECTOR is array (INTEGER RANGE <> ) of WORD;

    function WIRED_OR(SOURCES: BIT_VECTOR) return bit;
    function WIRED_OR(SOURCES: WORD_VECTOR) return WORD;

        component CLOCK_GEN  -- Synopsys TRANSLATE_OFF
        generic (ICLK_DEL, OCLK_DEL : Time);  - Synopsys TRANSLATE_ON
        port( ICLK : buffer BIT; OCLK : buffer BIT);
        end component;

        component BTS4
        port( A: in  BIT; E: in  BIT; Z: out wired_or BIT);
        end component;

        component BTS4_BANK
        port( IN_WORD : in  WORD; ENABLE  : in  BIT;
            OUT_WORD: out wired_or WORD);
        end component;

        component PAR_IN_SER_OUT
        port (DATA: in  WORD; LOAD: in  bit; CLOCK: in  bit;
            NINTO: out bit; O: out bit);
        end component;

        component SER_IN_PAR_OUT
        port (I: in  BIT; CLOCK: in  bit; NINTI: buffer bit;
            OUTPUT: out WORD);
        end component;

        component UART  -- Synopsys TRANSLATE_OFF
        generic (Clk_Per,ODel,INDel,INTDel: Time := 0 ns);  -- Synopsys TRANSLATE_ON
        port ( DATA: inout WORD; I: in BIT; LOAD: in BIT; READ: in BIT;
            O: out BIT; NINTO: out BIT; NINTI: buffer BIT);
        end component;

end TSL_Package;
```

Figure 25. Project Package for UART Design

The bidirectional data port of the UART is the first unit to be designed and requires the use of a three-state element that is drawn from a technology library. Figure 26 shows the code to implement the three-state cell. This cell is then used within a bank (whose width is determined by the WORD_LENGTH constant) that is created via the generate statement. The model of the primitive cell has been included to show how a simple model can be created.

Note, however, that error checking for three-state contention is not possible with this model. Because the cell is a library primitive, we can turn off synthesis translation around this component (via "— Synopsys TRANSLATE_OFF" interpreted comment). The resultant network after synthesis will simply reference the three-state driver network.

```
-- Synopsys TRANSLATE_OFF
use work.TSL_Package.all;
    entity BTS4 is
        port( A : in  BIT;            -- data input
        E : in BIT;            -- enable
        Z : out wired_or BIT);    -- resolved bit data output
    end BTS4;

architecture BTS4_HDL of BTS4 is
    begin
        process (A,E)
        begin
            if (E = '0') then Z <= '0'; -- 3stated state is zero here!!
            else Z <= A;
            end if;
        end process;
    end BTS4_HDL;
-- Synopsys TRANSLATE_ON

use work.TSL_Package.all;
    entity BTS4_BANK is
        port( IN_WORD : in  WORD; ENABLE  : in  BIT;
            OUT_WORD: out WORD);
    end BTS4_BANK;

architecture BTS4_BANK_RTL of BTS4_BANK is
    begin
        GEN_BANK : for I in WORD_LENGTH-1 downto 0  generate
            U1 : BTS4 port map(IN_WORD(I), ENABLE, OUT_WORD(I));
        end generate;
    end BTS4_BANK_RTL;
```

Figure 26. Bidirectional Modeling

Clock generation is not a generally synthesizable part of the overall design. For this example, a special entity is used that generates both the input and output clocks of the system. Again, for illustrative purposes, the simulation model portions of this design are surrounded by synthesis interpreted comments to control the processing of the model. The outcome of the synthesis process for the clock network is the instantiation of the CLOCK_GEN entity. The code for clock generation is shown in figure 27.

```
entity CLOCK_GEN is  -- Synopsys TRANSLATE_OFF
    generic (ICLK_DEL, OCLK_DEL : Time := 0 ns);  -- Synopsys TRANSLATE_ON
    port( ICLK : buffer BIT; OCLK : buffer BIT);
end CLOCK_GEN;

architecture CLOCK_GEN_RTL of CLOCK_GEN is
        signal tmp_ICLK, tmp_OCLK : BIT;
    begin  -- Synopsys TRANSLATE_OFF
        process(tmp_ICLK,tmp_OCLK) begin
            tmp_ICLK <= not tmp_ICLK after ICLK_DEL/4;
            tmp_OCLK <= not tmp_OCLK after OCLK_DEL;
    end process;
    ICLK <= tmp_ICLK;
    OCLK <= tmp_OCLK;  -- Synopsys TRANSLATE_ON
end CLOCK_GEN_RTL;
```

Figure 27. Clock Generation Simulation Code

The parallel-in serial-out block shown in figure 28 is based on a
WORD_LENGTH +1 length shift register. The shift register is loaded with the
input data word and an extra '1' at the top bit. As the word is shifted down, it
is zero filled. When the top WORD_LENGTH bits are all zero, shifting is then
complete and a completion signal is sent. Note that the completion signal
generated is for this module and not the final signal sent to the UART output.

```
use work.TSL_Package.all;
entity PAR_IN_SER_OUT is
    port (DATA  : in  WORD; LOAD  : in bit; CLOCK : in  bit;  NINTO : out bit; O  : out bit);
end PAR_IN_SER_OUT;

architecture PAR_IN_SER_OUT_RTL of PAR_IN_SER_OUT is
    signal OREG, NEXT_OREG : bit_vector (WORD_LENGTH downto 0);
begin  -- if LOAD then parallel load, else store next state
    SEQ_PAR_IN : process(LOAD,CLOCK,DATA,NEXT_OREG)
    begin
        if (LOAD = bit' ('1')) then OREG <= (bit' ('1') & DATA);
        elsif (CLOCK'event and CLOCK = '1') then OREG <= NEXT_OREG;
        end if;
    end process;

    COM_PAR_IN : process(OREG)  -- compute next state, and the completion signal
        variable go : bit;
    begin
        go := bit' ('0');
        for I in WORD_LENGTH downto 1 loop
            go := go or OREG(I);
        end loop;
        NINTO <= go;
        NEXT_OREG <= bit' ('0') & OREG(WORD_LENGTH downto 1); O <= OREG(0);
    end process;

end PAR_IN_SER_OUT_RTL;
```

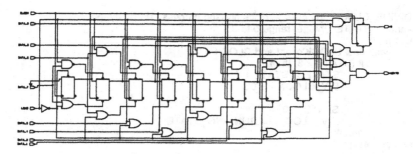

Figure 28. Code and Schematic for Parallel to Serial Conversion

The schematic for the synthesized model PAR_IN_SER_OUT is shown in figure 28. Note that the implementation has, in this case, been very tightly controlled by the explicit use of a shift register based description.

The serial-in parallel-out operation performed in the last subunit, SER_IN_PAR_OUT, uses a shift register architecture similar to the parallel-in module. This module receives a clock from the CLOCK_GEN module. The clock runs generating edges at four times the speed of the desired input rate. The clock is used as the basis for sampling, and then divided and used for shifting operations.

Embedded in the SER_IN_PAR_OUT model is a state machine that watches for the low transition of the data input I and makes sure that the input is low for two additional clock ticks before sampling begins. An enumerated type has been declared for the state machine's state values. The completion signal, NINTI, is also controlled within the state machine process. The schematic for the synthesized serial-in to parallel-out converter is shown in figure 29.

```
use work.TSL_Package.all;
    entity SER_IN_PAR_OUT is
        port (I: in  bit; CLOCK: in  bit; NINTI: buffer bit; OUTPUT: out WORD);
    end SER_IN_PAR_OUT;

    architecture SER_IN_PAR_OUT_RTL of SER_IN_PAR_OUT is
        type I_STATE is (RESET_STATE, I_LOW_1, I_LOW_2, I_LOW_3, RUNNING);
            -- enum type for FSM
        signal STATE, NEXT_STATE : I_STATE;
        signal IREG : bit_vector (WORD_LENGTH downto 0);
        signal DIV_CLOCK, RESET, RUNNING_STATE : bit;
```

```
begin
DIVIDE_CLOCK : process  -- clock dividing process - divide by four
variable STATE : integer range 0 to 3;
begin
    wait until CLOCK'event and CLOCK = '1';
        if (STATE < 3) then
            STATE := STATE + 1;
            DIV_CLOCK <= '0';
        else
            STATE := 0;
            DIV_CLOCK <= RUNNING_STATE;
        end if;
end process;

SHIFT_IN : process(DIV_CLOCK) -- shift register process with asynch clear to 0...01
begin
    if (RESET = '1') then
        IREG(WORD_LENGTH downto 1) <= "00000000"; IREG(0) <= '1';
    elsif (DIV_CLOCK'event and DIV_CLOCK = '1') then
        IREG <= IREG(WORD_LENGTH-1 downto 0) & I;
    end if;
end process;

OUTPUT <= IREG(WORD_LENGTH-1 downto 0); -- output get low word_length bits

SAVE_STATE : process begin -- state machine state register declaration
    wait until CLOCK'event and CLOCK = '1';
        STATE <= NEXT_STATE;
end process;

FIND_START : process(I,STATE,IREG)  -- find low I bit (low for half period = 2 more ticks
begin
    case STATE is
        when RESET_STATE =>
            if (I = '0') then NEXT_STATE <= I_LOW_1;
            else          NEXT_STATE <= RESET_STATE;
            end if;
            NINTI <= '1';
        when I_LOW_1 =>
            if (I = '0') then NEXT_STATE <= I_LOW_2;
            else          NEXT_STATE <= RESET_STATE;
            end if;
            NINTI <= '1';
        when I_LOW_2 =>
            if (I = '0') then NEXT_STATE <= I_LOW_3;
            else          NEXT_STATE <= RESET_STATE;
            end if;
            NINTI <= '1';
        when I_LOW_3 =>
            if (I = '0') then NEXT_STATE <= RUNNING;
            else          NEXT_STATE <= RESET_STATE;
            end if;
            NINTI <= '1';
```

```
            when RUNNING =>
                if (IREG(WORD_LENGTH) = '0') then
                    NEXT_STATE <= RUNNING;
                    NINTI <= '1';
                else
                    NEXT_STATE <= RESET_STATE;
                    NINTI <= '0';
                    end if;
        end case;

        if (STATE = RUNNING) then
            RESET <= '0';
            RUNNING_STATE <= '1';
        else
            RESET <= '1';
            RUNNING_STATE <= '0';
        end if;
    end process;

end SER_IN_PAR_OUT_RTL;
```

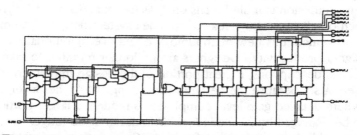

Figure 29. Code and Schematic for Serial-to-Parallel Converter

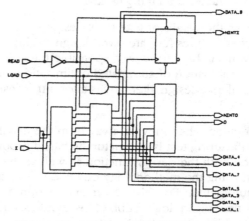

Figure 30. Top Level UART Schematic

5.7 Closing Remarks

The migration to VHDL is gaining momentum. Many benefits await the adopter of VHDL as the language of choice in an HDL synthesis design methodology. Because the language is a standard, designs are portable to alternate design environments. Also, there are numerous sources of models in VHDL which can be leveraged for new designs. Using a standard permits model communication between vendors, contractors, and tool sets. These factors will make assembling the right set of tools for the job easier than ever before.

The descriptive power of VHDL allows the designer to use one language for simulation and synthesis. Thus, you need to learn only one language to complete the entire design task. Using this synthesis methodology allows compliance with industry and defense standard VHDL documentation requirements. This will aid in the long term lifecycle costs of an electronic design.

Why will the HDL synthesis methodology be the norm of tomorrow? The answer to this question is multidimensional, as the benefits of HDL synthesis design touch on all of the key areas of complex system design. The most significant improvement this new methodology offers over traditional methodologies is designer productivity. Designers are no longer required to manually enter gate level descriptions, a great portion of the design implementation time is saved. Also, because the designer is working at a higher level of abstraction where detailed gate level complexity is hidden, a larger portion of a design can be managed by an individual. The designer no longer need familiarize himself with the peculiarities of each ASIC vendor's libraries; technology specific aspects of design are automatic. Furthermore, moving up to the RTL level allows more of the designers thinking to be at a conceptual and technology independent level, rather than the gate level.

Moving away from the details of gate level design does remove the importance of the established metrics for gate level design quality: speed, area and test. The designer will still be ultimately responsible for design quality along these axes, but the combination of automated optimization and the capability to quickly explore multiple design alternatives allows the designer to achieve higher quality final results.

Another consideration when moving away from the details of gate level design is test. Manufacturing test is a structurally based problem whose complexity depends upon both the design function as well as the exact gate level structure. However, using the methodology outlined in this chapter, structural implementation is automated. For this reason, automation of the test process is a critical issue. A new technology, dubbed test synthesis, is emerging as a solution to the test problem for synthesized logic.

The key considerations in a test synthesis implementation are the design for test (DFT) and automatic test pattern generation (ATPG). Both of these issues are critical to testing a synthesized design. DFT ensures that the design will be inherently testable. This implies the synthesis of irredundant networks and the automatic inclusion of necessary structures (e.g., scan chains), and a follow-up optimization of the network. The follow-up optimization is essential because of the change in speed and area characteristics of the design with DFT structures in place. Once these changes are made, of course the documentation for the design needs to be regenerated (schematics and HDL); most test synthesis tools can do this automatically. ATPG on a testable network is a more tractable problem than test generation on an arbitrary network. Beginning with a testable design, high fault coverage can be consistently achieved. Test synthesis completes the picture of gate level implementation issues, leaving designers free to design at the HDL level.

The benefits cited above all come as a by-product of adopting the revolutionary new technology of HDL synthesis. The true beauty of this revolution is that it represents only an evolutionary change to most existing gate level methodologies. Traditionally, a designer spent a significant amount of time entering the schematic drawing for his design. With HDL synthesis, that portion of the design cycle is automated. The design sign-off path after netlist creation does not change. The incremental change to the design flow is the creation of a synthesizable HDL description of the design. This description is easier to create than a netlist because of the higher level of abstraction used to generate it and can be used as a more simulation efficient means of verifying the chip at the board level.

To summarize, the steps of the HDL synthesis methodology are:

1. Composition of the synthesizable HDL model
2. Validate the behavior of the high-level model
3. Synthesize (translation + optimization) to gate level
4. Verify the design with the technology specific parameters

HDL synthesis methodologies will become the norm both because of the complexity of the present and next generation ASICs with accompanying reduced time to market pressures, as well as the increased design quality obtained through designer control at the higher levels of abstraction and automation of the detailed design.

Chapter 6
Modeling of Standard
Component Libraries

Gabe Moretti

RISC International Systems Corp.
San Jose, CA

This chapter deals with the use of VHDL in an area where methodology is still in a state of flux. The intent is to describe the more important and timely issues of model development in VHDL and to further stimulate discussion and the exchange of ideas contributing to the emerging standard modeling methodology. The interface protocol and semantics of models are characteristics most relevant to a standard making body. Since VHDL is an IEEE standard, the standard modeling methodology must be both predictable enough to insure longevity and flexible enough to allow competition through flavors of implementation. Strict coding guidelines for VHDL will be counter-productive. Interoperability, not similar internal structure, is the goal of the successful standard.

The power and flexibility of VHDL could be viewed as more of an obstacle than an aid to establishing a standard modeling methodology. A limited tool forces a discipline on its user which is inherent with its shortcomings. But VHDL's richness allows a significant degree of freedom in the modeling methodology employed, greater latitude in the level of abstraction of the whole model, and its parts.

Building interoperable models means that models from different sources must not only communicate with each other through some common protocol, but also that their level of abstraction is similar. The simulation of a given circuit is successful not just because syntactical errors have been avoided, but because it produces meaningful results which help in the understanding, development, and documentation of the design.

As with every other digital logic modeling system, the acceptance of VHDL as a modeling language depends heavily on the availability of usable models, not just the power of the simulator. Model development is a task which requires the investment of costly resources for a significant period of time. Yet such a task cannot be an overhead activity to every digital design project. Product development schedules require models be built before and shared during the design development cycle. If VHDL, no matter how powerful, injects more overhead in the development cycle than is already present, it will not be widely adopted.

Without an agreed upon set of guidelines, VHDL models of similar parts may be as different from each other as models written in different simulation languages. The danger is that while simulator specific models provide an outward indication of incompatibility by looking different, VHDL models will appear similar, and therefore invite the user to believe that interoperability is implied. The tendency may be to trust the results of the simulation, even when incorrect, thus creating costly mistakes. It is necessary, therefore, to develop a usage standard in addition to the syntax and semantics of VHDL. We will then avoid, or at least isolate in easily identifiable areas, discrepancies in the semantics and levels of abstraction in models.

1. Structure of Model Libraries

In defining a methodology for the development of interoperable models one must begin by recognizing that the most important attributes of models are their ease of use. When a user receives a model, he or she must have access to the complete set of definitions and utilities which support the portability of the model. At the same time, one must be mindful of the fact that storage space is neither infinite nor infinitely cheap, and that too much data is just as dangerous as not enough.

With the exception of special situations, users need only a portion of the source code in order to know how to use a model. Therefore, the organization of information in VHDL libraries and packages must be given some thought before one gets too involved in the details of methodology specifications.

The intent of interoperable VHDL models would indicate that libraries of such models would be distributed in source form, at least until a standard intermediate form is accepted. Yet one can envision a vendor providing already analyzed models for specific VHDL systems, in order to protect certain intellectual property rights. Even in the latter case, the end user must have access to portions of the source code, in order to understand how to use the model in his or her design description.

The minimum amount of information required in VHDL source form is the model's **entity** declaration and portions of the referenced **packages**. **Package** portions needed are those that define interface **subprograms, type conversion** algorithms, and globally accessible **constants** used to control the execution flow of the models.

A library of models generally share a number of utility subprograms. An efficient way of grouping VHDL code is to divide it into two **packages**. One **package** contains **type** definitions, **constant** definitions, and **subprograms** which are required to interface the models in the library to other VHDL **design units**. The other **package** contains the utility **subprograms** necessary to perform timing calculations, **type** conversions on variables, logic operations on registers and memories, and the like. In this chapter the former **package** is called SYMPACK, while the latter is called UTILPACK.

The ability to overload language elements is a source of elegant solutions to a number of problems. Using such a technique in portable **packages** is very useful but requires complete documentation.

Overloading of predefined names should in general, not be done by end users and should be well documented by **package** developers. The danger is that a user might inadvertently **overload** either a **type** or an operator already overloaded in a portable **package**. Such an occurrence is not likely to generate a compilation error. Its impact will be difficult to isolate and debug.

Standard modeling guidelines must discourage operator **overloading** not accompanied by explicit documentation.

Developers of portable **packages** should also carefully choose identifiers for **type** definitions and **subprograms** which are unique during common use of the **packages**. The **packages** provided with this chapter are an example of such a methodology.

A methodology proposed here is the use of only one **architecture** per **entity** declaration in order to minimize the overhead required to use the library. Otherwise, there should be a unique **configuration** for every entity/architecture pair that is parametized. In the **component specification** the user deals with either the **entity** name or the **configuration** name. The name of the **configuration** should be a variant of the **entity** name obtained by applying a fixed transformation as described in the library documentation. The naming should apply to all of the models in the library.

The approach taken in developing the structure of the models is one which minimizes the implementation of algorithms at each layer of abstraction. Such a tactic allows the greatest possible degree of freedom to the user by avoiding the implementation of decisions which would constrain lower levels of abstractions needlessly. This will be particularly useful in the handling of error conditions and timing parameters and is evident within the examples given with this chapter. Deferring implementation until the correct layer of abstraction is reached is a powerful method to assure interoperability.

2. Relevant Issues in Logic Simulation

The first step in minimizing coding is to recognize that the functionality of a standard component should be described independently from its I.C. packaging. The model should be designed for easy encapsulation in a physical package representation which supports engineering tools. The functionality of a part should then be mappable onto the technology used for an implementation. The model of functionality of a part should not incorporate assumptions regarding the interconnect technology. The collection of traces on a PCB layer can be regarded as a model in its own rights, thus allowing the engineer to deal with the physical properties of the layer in a localized manner. These issues are relevant to the methodology used in simulating signal propagation delay in the circuit, as well as the naming and ordering of ports in the entity declaration.

Since the introduction of VHDL the first discussion point among the vendor and the user communities has been the choice of a strength/state value set for modeling. Due to the existence of well established logic simulators, the introduction of VHDL has and will continue to be influenced by the need to provide an orderly migration path between a traditional simulator and VHDL. Therefore an interoperable model must be able to interact with a number of strength/state value sets. The methodology must support the migration path, not make it more difficult. The strength/state value set is one of the areas in which CAE companies resist standardization in order to achieve a competitive advantage through product differentiation. A common strength/state value set must be supported but a methodology to deal with mixed strength/state models must also be developed.

Finally, when planning a library of models one must take into consideration the impact that modeling methodology will have on the storage requirements and the ease of use of the library. Standard parts, in particular, have a number of speed and pinout variations associated with the same logic behavior. A methodology must be used in dealing with parts which minimizes the need for code duplication and recompilation in order to modify its timing, for example.

The remainder of this chapter will explore a methodology for the design and development of models of integrated circuits in VHDL which satisfy the following requirements:

- Usable at different stages of the design process with maximum cost effectiveness

- Independent from IC packaging, i.e. pinouts and functional duplication

- Independent from the Strength/state values set supported by the target simulator

- Independent from specific propagation delay calculations and technology related timing values

- Independent from the routing and layout functions

3. Layers of Abstraction

Logic simulation can be used at different stages in the design cycle. Yet practically all presently available CAE systems support logic simulation with just one set of models. The result is that the model will provide an effective engineering tool with reasonable user interface at the stage in the design cycle for which it was originally intended, and may be awkward to use or have too much execution overhead at other stages.

The requirements for a model used in logic synthesis, circuit design, or for post-routing verification vary greatly. Until the acceptance of VHDL, which supports modeling at many different layers of abstraction, the requirements were met by various vendors with multiple languages and libraries.

To highlight the problems and issues, one needs to study the design methodology, understand the needs at various design stages, and provide for them in the appropriate model. The layers of design abstraction and their needs are exposed next.

When VHDL is used as a system architecture description language, the component should model the desired behavior of the part without handling error conditions, either as input to the model or as an output from it. Signal propagation should be handled in zero delay fashion. The idea is to represent the functional aspect of the abstract part. This style of modeling is well suited to support feasibility studies in the early stages of a design, as well as to fully describe the functional behavior of a design to a synthesis tool when the logic design is completed.

Note that although models written for zero delay are common and very useful, they are difficult to debug and use properly in some implementations. Most debug environments for VHDL only allow analysis in a macro (one time unit) setting and not delta (zero time unit) delay setup. Check your implementation debug system for support of your needs.

Models are also useful when simulating actual circuitry in order to validate its correctness. They provide, in addition to correct functionality, error detection and generation, as well as signal propagation delays and timing checks. The model supports the evaluation of the operating range of the circuit by providing a way to derate timing based on operating conditions. Signal propagation delays due to interconnections between models are also taken into account when evaluating the final circuit layout.

Since too much data is just as confusing as not enough data, models will many times have a mechanism to select the desired layer of abstraction for a given simulation. One possible implementation is to have many **architectures** for each VHDL **entity** (one for each layer of abstraction) and allow the user to select the required one at **component** specification time. Such a strategy, although elegantly supported by the language, presents two significant drawbacks.

First, the user must choose the required functionality by explicitly naming the **architecture** in the instantiation specification. This method offers no built in validation that the correct level of abstraction has been picked, as long as an **architecture** with that name exists in the library. Such strategy also requires code duplication among the **architectures**, since lower levels of abstraction still need all of the processing required by the higher level. This fact would complicate the maintenance of the models and significantly increase the size of the libraries.

Incorporating the support for all the layers of abstraction into one **architecture,** and providing the user with a mechanism to choose which portion of the **architecture** to execute is preferable. Such choice could be made through the use of VHDL **generics,** but the use of **deferred constants** is a more efficient and elegant way to provide control to the end user. The use of **deffered constants** allows changing their values in a central **package body**. This strategy makes it possible to change the **constant** value in a localized manner without requiring that all **design units** present in the library be re-analyzed. Only the central **package body** containing the **constant** default value needs to be re-analyzed after the changes.

The ability to choose the layer of abstraction to be executed in a model requires logic flags and conversion functions which must be readily available in source code to users. The **package** SYMPACK included in with this chapter is an example of an implementation of this methodology.

```
Package sympack is
  constant Logic_Error_Handling: boolean;
  constant Tpd_Handling: boolean;
  --
  -- more definitions to be shown later
  --
end sympack;
  --
Package body sympack is
  constant Logic_Error_Handling: boolean: = FALSE;
  constant Tpd_Handling: boolean: = FALSE;
  --
  -- more specifications to be shown later
  --
end sympack;
  --
```

FIGURE 1 Model Execution Control with Package

A fragment of the **package**, contained in Figure 1, shows how to control the execution of models as a function of the type of simulation required. The assumption is made that the minimum functionality required from each model is logic simulation with zero delay and no error handling. Flow control flags are defined in the **package** and given their values in the **package body**. By setting them in the body, the user can easily modify the values and thus control the level of model abstraction to be used as frequently as necessary.

```
architecture a_name of an_entity is
-- declarations
begin
  process
  begin
--
-- calculation of output signal values for good circuit behavior
--
    if Logic_Error_Handling then
--
-- Identification of input error conditions generation of error conditions
-- reporting of error conditions
--
    end if;
    if Tpd_Handling then
--
-- calculation of propagation delays and test of relevant timing parameters.
-- Generation of error messages if timing parameters have been violated
--
    end if;
--
-- Assignment of calculated signal values to appropriate ports after either
-- calculated or default delays
--
  end process;
end a_name;
```

FIGURE 2 Behavioral Architecture Template

Using this methodology the contents of an **architecture** design unit will have the structure shown in Figure 2. Such structure describes the part in decreasing degrees of abstraction, mimics the organization of a data sheet to provide the required documentation of the part, and decreases maintenance costs by providing a regular structure of loosely connected code segments. Note also that VHDL **generate** statements could perform a similar function when the selected code is actually concurrent.

4. Independence from Physical Packaging

The structure and semantics of the VHDL code contained in the **entity** declaration of standard parts are of primary importance. In most cases it constitutes the only portion of the model visible to the end user. The need to protect the engineering rights of semiconductor and modeling companies will be satisfied by strict restrictions on the availability of source code of **architecture** design units as well as **package body** design units which deal with technology features. In addition, as the number of verified parts increases, the use of these models will become more widespread. Therefore a designer must be able to easily infer the proper use of the model from the **entity** design unit and the **package body** associated with the library.

Given an existing library of models, a user can configure a model in his or her design through the **component** declaration with **generic** and **port** interface lists. The position of a formal parameter within the **generic** interface list can be quite arbitrary, so it is important that the **generic** variables be well documented. The use of short, cryptic names saves typing, but may cause confusion.

If the user is in an environment where models originate from more than two sources, then **component instantiations** should use named instead of positional association. In this way the correlation is explicit, instead of the implicit binding provided by positional association.

Each formal generic of a **component** declaration should have an associated default value in the **entity** declaration. This facilitates the use of the model before all values for **generics** have been determined. The practice assumes increased importance when the user does not have access to the source code of the **architecture** body of the module, since a "reasonable" guess at a default value could have an unwanted impact on code execution.

The order of formal port parameters and their actual names is a more complex issue to standardize. The semiconductor companies have not themselves used any standard naming convention in data sheets. There are often differences in signals names even with pin-for-pin replaceable parts. The use of special characters, like slashes, starting a signal name with a number, and using the overbar as an indication of an active low signal, are all regrettable obstacles to the standardization of signal names in a computer processible format. Even the C.A.E. tools themselves complicate the problem because they often employ unique conventions and characters.

The **port declaration** in the **entity** design unit describes a **signal** interface between the model and the rest of the system. This interface must be at the functional level and not related to any specific integrated circuit packaging technology. Ignoring the actual packaging provides more freedom of choice, following the engineering process where **signals** are not associated with actual pins until layout.

The packaging of a logic function component can be elegantly achieved in VHDL by instantiating the functional model within a design unit which represents the to be used. The resulting entity can now be used for layout without imposing any undue restrictions on the previous engineering steps.

Most data books published after 1986 contain the logic symbol of the part according to the ANSI/IEEE Std-91 guidelines. Developers of VHDL models should use such logic symbol standards as the guide to establish the order in which ports are listed. The order should begin with the signal at the top left hand side of the symbol and proceed counter clock wise until the top right hand side signal name has been described in the port interface list.

Note that symbols drawn according to this standard will not show any functional signal on the top or bottom side of the drawing. Furthermore, since the input signals are drawn on the left hand side, and the output signals on the right, the ports will be ordered according to their mode. In addition, the ANSI/IEEE Std-91 groups signals according to their functional semantics, thus the sequence is not arbitrary.

By following this technique, a simple mapping between the logic symbol of a part and its simulation model is achieved. Such a mapping could be used in the machine generation of a VHDL entity interface from a symbol editor or EDIF description, or vice-a-versa

Figure 3 shows an example logic symbol with the corresponding **VHDL port interface list**. Note that the signal names had to be modified from the original names to ones which are allowed identifiers in VHDL. This is a common problem which will have to be addressed until symbols begin to follow language naming requirements.

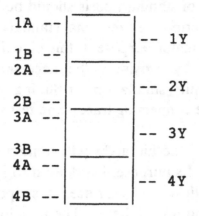

```
entity STD_86 is
port(A1, B1, A2, B2, A3, B3, A4, B4: in logic7;
    Y4, Y3, Y2, Y1: out logic7);
end STD_86;
```

FIGURE 3 Logic Symbol and Corresponding Entity

A significant portion of MSI and SSI parts have more than one equivalent function packaged in one device. Handling of this type of standard part has created dilemmas in the past whenever a CAE company wanted to establish a link between a schematics editor and a simulator. Both system design and circuit design deal with functional elements, not with packages, and therefore an engineer does not want to draw two flip-flops when only one is needed. On the other hand, during layout the flip-flop must be part of a physical package, so a model of the actual standard part with multiple functional elements is also necessary.

Most CAE companies have resolved the issue by providing both types of models, one of the individual functional device, and another of the standard part. VHDL provides syntactical support to this strategy through the component instantiation mechanism. Models of this class of standard parts should be hierarchical. The top level should describe the functional standard part as packaged by the vendor. The major purpose of this model is to describe the pinout of the part by providing the necessary connections to support final circuit simulation including load and drive characteristics of the connecting traces, and test vector generation.

The lower level of the hierarchy will contain the model of one functional element. Its purpose, in addition to support for system design and schematic entry interface, is support for functional circuit simulation and for synthesis. The functional model can be instantiated as many times as necessary into the circuit level model to provide complete support to the final system simulation and interface to the layout function. For example the model for an '86 TTL part would instantiate the functional model (a two input exclusive or gate) four times.

Although the use of VHDL descriptions for actual layout information is not widely supported, the description of the actual positioning and interconnect of the parts in VHDL is a necessary step in the design process. Therefore description of the pinout of a part within a given package is required. A flexible approach to satisfy this requirement is to describe an IC package as a design unit which instantiates the needed logic function. See Figure 4 for an example using the SN54LS86A in the J package and Figure 5 for an example using the same device in the FK package.

```
entity DIP_86 is
  port (P1, P2: in logic8;
      P3: out logic8;
      P4, P5: in logic8;
      P6: out logic8;
      GND: linkage logic8;
      P8: out logic8;
      P9, P10: in logic8;
      P11: out logic8;
      P12, P13: in logic8;
      VCC: linkage logic8);
end DIP_86;
architecture i20x09 of DIP_86 is
  component STD_86
    generic (part_type: string);
    port (A1,B1,A2,B2,A3,B3,A4,B4: in logic7;
        Y4,Y3,Y2,Y1: out logic7);
-- generic and port clause
  end component;
begin
  U1: STD_86
    generic map (part_type = > "SN54LS86A")
port map (A1 = > P1, B1 = > P2, Y1 = > P3, A2 = > P4,
    B2 = > P5, Y2 = > P6, Y3 = > P8, A3 = > P9,
    B3 = > P10, Y4 = > P11, A4 = > P12, B4 = > P13);
end;
```

FIGURE 4 SN54LS86A J Package Pinout

The package provided with the components library will define VCC, GND, NC, and all other pseudo signals so that the designer is freed from having to designate these connections explicitly.

Notice that the pseudo-signals GND and VCC are designated of mode **linkage.** Following this methodology will assure that the pins will not be accidentally connected to actual **signals.** An error will be reported before any time is wasted in simulation. A **signal** mode can be used to identify possible misconnection. For example, if the output of a device is known to be totem pole, the output port should use **mode** buffer instead of out or an unresolved type. Doing this will insure that if more than one driver is associated with the **signal** an error message will be output before simulation starts.

The strategy of encapsulating a functional model in a specific IC package description for layout verification offers an additional advantage. IC package models are connected via **signals** to their functional models. Delays due to trace lengths and fanout can be computed and assigned when the **signals** are connected; instead of the traditional approach of computing input delays inside the model.

This strategy offers a number of advantages over the traditional approach of handling load dependent delays and fanout independently of the functional models. A change in the derating equations is easier to implement, since the contribution of the factors under discussion is modularized and handled separately from internal device delays. The derating equation can be included in a **package body** which is distributed in intermediate form only, to protect proprietary information. Furthermore, derating equations vary from vendor to vendor. It is difficult to standardize the number and type of terms required by the equations. To provide support to all the methods inside models of standard parts would require not only even more **generic variables**, but also a decision tree which will slow down simulation and increase maintenance costs.

```
entity FK_86 is
 port (NC_0: linkage logic7;
      P2, P3: in logic7;
      P4: out logic7;
      NC_1: linkage logic7;
      P6: in logic7;
      NC_2: linkage logic7;
      P8: in logic7;
      P9: out logic7;
      GND, NC_3: linkage logic7;
      P12: out logic7;
      P13, P14: in logic7;
      NC_4: linkage logic7;
      P16: out logic7;
      NC_5: linkage logic7;
      P18, P19: in logic7;
      VCC: linkage logic7);
end FK_86;
architecture i20x20 of FK_86 is
 component STD_86
   generic (part_type: string);
   port (A1,B1,A2,B2,A3,B3,A4,B4: in logic7;
         Y4,Y3,Y2,Y1: out logic7);
 end component;
begin
 U1: STD_86
   generic map (part_type = > "SN54LS86A")
   port map (A1 = > P2, B1 = > P3, Y1 = > P4, A2 = > P6,
 B2 = > P8, Y2 = > P9, Y3 = > P12, A3 = > P13,
 B3 = > P14, Y4 = > P16, A4 = > P18, B4 = > P19);
end;
```

FIGURE 5 SN54LS86A FK Package Pinout

5. Strength/Level Values Set Independence

VHDL allows user defined signal **types**. This freedom makes it possible to tailor a model to very specific needs, or even to mimic the behavior of another simulator in VHDL. Such freedom is a major obstacle to mixing models from various vendors in one simulation.

There are two major reasons for the proliferation of different **signal types** within VHDL. The first reason is that different layers of abstraction require different **signal types** as one transitions from boolean algebra to gate to electrical. The closer one gets to the transistor level the richer the set of values used to describe a state must be. For example, a 46 value set is required to accurately model a device at the switch level.

The second reason for the proliferation is more pragmatic. In an effort to preserve their engineering investments and to provide a transition path from an existing simulator to VHDL, vendors have built a VHDL front-end to their logic simulators. Such "hosted VHDL" systems still use the simulation engine of the existing simulator. Therefore only **signals** which have values supported by the simulation engine can be handled.

There are two strategies proposed to build interoperable models which are portable to various VHDL implementations. The first strategy is proposed by the VHDL Design Exchange Group (VDEG), a part of the VHDL Users' Group. VDEG has identified the minimum logic necessary for logic level (not switch) ASIC netlist modeling, using predefined cell models. The set contains four levels and no strengths; the levels are '0','1','X', and 'Z'. The 'X' value represents both unknown and uninitialized; the 'Z' is a representation of a non driving tri-state. Although this set is sufficient to describe the functional behavior of a circuit, it is not sufficient to accurately describe board designs, or I/O cells in gate arrays, or mixed technology designs in general.

To accurately simulate the interaction of a **signal** with another which has the same logic state but different drive characteristics one needs a minimum of seven levels: three states, two strengths, plus the floating tri-state level. The Electronic Industry Association (EIA) is developing a package which supports such a set augmented with the state 'U' for uninitialized. This state is essential in distinguishing between "unknown" and "uninitialized" for simulation verification. The SYMPACK **package** used in the example contains the equivalent types to the four, seven, and eight value sets with only minor modifications. The elements of **type** logic8 are ordered in such a manner that both the **type** logic7 and the **type** logic4 can be defined as **subtypes**.

Given the existance of a diverse number of strength/state value sets, one can use the power of VHDL to follow a strategy which isolates the model from its environment. This is done by performing the required **type** conversion in the association list of a **port map**. This requires that all **type** conversion functions be provided in the library support **package**. The **package** SYMPACK shows examples of what is required.

Portions of conversion functions are shown in Figure 6, using technology descriptions borrowed from the Zycad public domain **package**. The appeal of the VDEG proposal is that the values of the signals are familiar to every user, since their names are the same in all VHDL systems. But, models developed solely according to the VDEG guidelines have less functionality and require **signal type** transformation in order to be used with other state systems. The fact that the values of a **signal** set are a subset of another has no meaning in VHDL unless either the former's **type** is defined as a **subtype** of the latter, or both sets are explicitly defined as **types** and a conversion function between the two is provided.

By using the **named association** in the **port map**, a conversion function can be indicated at **component specification** time, resulting in an efficient solution to interfacing models with differing **signal types**. Of course this strategy requires that the target VHDL simulation system support user defined **signal types**. Within an individual model it is only necessary to calculate the level of a signal, not its strength. The drive and sense functions in **SYMPACK** demostrate that strength is a function of technology, while the level is a result of circuit logic. Such an approach simplifies the model. The result is a model which is portable and maintainable.

Using functions called **sense** and **ttl_drive** for input and output ports respectively, Figure 7 shows how a standard part implemented in four level logic can be used in a simulation which employs a seven level logic value set. As can be seen from the examples provided in the **SYMPACK package**, the sense and drive functions can be written in a manner which makes them very efficient during execution, since they are the equivalent of a table lookup.

A **type** transformation which goes from one set to another with less possible values must collapse some portion of the larger set semantics. The process is equivalent to increasing the level of abstraction in the design description. For example moving from seven to four value set means that a totem pole output becomes indistinguishable from an open collector output. Conversely, a transformation to a larger set of values is equivalent to decreasing the level of abstraction. To do so, and still maintain at least the same degree of precision as the source values set, requires additional information.

```
package sympack is
  type logic8 is ('U','X','0','1','Z','W','L','H');
  subtype logic7 is logic8 range 'X' to 'H';
  subtype logic4 is logic8 range 'X' to 'Z';
  type tec_type is (ttl, cmos, ecl, pmos, nmos, jam_latch,
               bus_keeper, open_collector, open_drain);
  type sense724 is array (logic7'low to logic7'high) of logic4;
  type vector427 is array (logic4'low to logic4'high) of logic7;
  type table427 is array (tec_type'low to tec_type'high) of vector427;
  constant rsense: sense724: = ('X'',''0'',''1'',''Z'',''X'',''0'',''1'');
  constant drive_tab:table427: =
      ((('X','0','1','Z'),  -- ttl
       ('X','0','1','Z'),  -- cmos
       ('X','L','1','Z'),  -- ecl
       ('X','Z','1','X'),  -- pmos
       ('X','0','H','Z'),  -- nmos
       ('W','L','H','X'),  -- jam latch
       ('W','Z','H','X'),  -- bus keeper
       ('X','0','Z','X'),  -- open collector
       ('X','Z','1','X'));  -- open drain
  function sense (s1:logic7) return logic4;
  function ttl_drive (s1:logic4) return logic7;
-- rest of package
end sympack;
package body sympack is
  function sense (s1:logic7) return logic4 is
  begin
    return rsense(s1);
  end sense;
  function ttl_drive (s1:logic4) return logic7 is
  begin
    return drive_tab(ttl)(s1);
  end drive;
-- rest of code
end sympack;
```

FIGURE 6 Conversion Function

In the example used previously the **type** conversion function must be able to differentiate a totem pole output, which at logic level high would have a value of '1', from an open-collector output, which would receive the value 'Z' when the four value set is mapped to the seven value set.

```
entity STD_86 is
-- generic(part_type:string);
 port (A1,B1,A2,B2,A3,B3,A4,B4: in logic7;
     Y4,Y3,Y2,Y1: out logic7);
 constant t1,t2,t3,t4: tp_values:= (0 ns, 0 ns, 0 ns);
end STD_86;
architecture A_STD86 of STD_86 is
 component EXOR2
  generic (d1,d3,d2,d4: tp_values);
  port ( A, B: in logic4; Y: out logic4);
 end component;
begin
 U1:EXOR2 -- one of four instantiations
  generic map It1,t3,t2,t4)
  port map (A = > sense(A1), B = > sense(B1), ttl_drive(Y) = > Y1);
end A_STD86;
```

FIGURE 7 Use of Conversion Functions

6. Independence from Timing Parameter Values

Two important characteristics of interoperable models are flexibility and modularity. If a model is developed with the goal to optimize these two properties, the result is either a model which can interface to other models, or which can be modified in a cost effective manner to satisfy the new requirements. One of the obvious ways to modularize a model is to separate the handling of the logic functions from the handling of time calculations. A number of proposals to achieve such modularity have already been put forward.

The EIA, as well as a number of VHDL vendors, have proposed packages which contain functions to implement pulse width checks, setup and hold, and propagation delay computations.

Generally, IC vendors provide a number of speed variations for each standard part. Gate array libraries have common functional cells but different speed characteristics and formulas for signal propagation delay calculation. Such commonality of functions must be recognized and exploited by modelers. Therefore, to minimize the size of part libraries, it is imperative to separate timing from logic and develop a methodology to provide the appropriate timing parameter values to the model.

Unfortunately, the information contained in the data books from semiconductor vendors is not in a standard form, and therefore some fundamental work must be done before the data can be used in a model. Design engineers generally prefer to have propagation delay values expressed in the form of minimum, typical, and maximum delay as a function of operating conditions. The example in Figure 8 defines a **type** which groups these values in the **package** SYMPACK.

---

type db_time **is** (min, typ, max);
type tp_values **is array** (db_time'low **to** db_time'high) **of** time;

FIGURE 8 Support of min/typ/max Timing with Types

Most vendors will frequently omit one or two values from the minimum/typical/maximum specification. Modelers should agree on a common way to derive the missing values. One strategy, introduced by HHB Systems for its CADAT simulator, is shown in Figure 9. The table gives the ratios between minimum, typical, and maximum values as a percentage of the typical value, for various fabrication technologies. Using the table, a modeler is able to calculate the value of the delay parameter omitted by the vendor. Although from time to time a particular vendor might disavow a particular value not specifically contained in the data book, the ratios have been verified with a number of IC vendors and provide the best guess default value.

At the present time some working groups within the IEEE are considering proposals which would utilize a data file as the vehicle to provide timing parameter values to the models. This approach solves some problems, notably verbose generic clauses, and limits the need for re-analysis when changes are made. But it generates more problems than it solves.

Family	Minimum	Maximum
TTL	35	190
FTTL	60	145
ASTTL	60	160
ALSTTL	35	155
HTTL	35	155
STTL	35	150
LSTTL	35	165
ECL	60	150
NMOS	50	150
PMOS	50	150
CMOS	35	200
HCMOS	35	200
HCTCMOS	40	170
FCTCMOS	50	160
ACCMOS	40	190
ACTCMOS	50	230

FIGURE 9 MIN/MAX Delays as a Percentage of Typical Delay Value

Interoperable models must use only **TEXTIO** facilities to read files in order to guarantee execution in all **VHDL** systems. In addition, the amount of timing data required to support a library of models that implements one data book is quite large. Only two implementation strategies are therefore feasible: one single file containing all of the data required by one library, or one data file per model in the library.

The use of the **TEXTIO package** has a significant negative impact on the execution speed of a simulation, a fact which can be easily measured when running a test bench during model validation. If the strategy used is one file per library, each model instance will, on the average, have to read half of the file in order to find the required data; an extremely inefficient approach.

Timing data will usually be read at VHDL elaboration time as part of a **constant** value specifications. The time required to do such a read operation when simulating a printed circuit board or gate array of average size and complexity is significant. The proposed approach requires each model instance to read its values. For a moderate size design of 40,000 gates this would require up to 40,000 file opens and accesses!

Furthermore, the use of data files which are loosely coupled with a model creates a version control problem, since there is no built in VHDL mechanism which will guarantee that the data file read to retrieve timing values for a given model is the most up-to-date one. Each installation will have to provide an acceptable mechanism for the safeguard of the official timing data files. Failing to do so will undermine the concept of leaf level module, since such modules now must be paired to their corresponding leaf level timing files. This will complicate the administration of the leaf level modules data bank.

A better solution is to use a system of explicilty declared **constants** in the **entity** declaration of each model. Such a system allows the modeler to fully document the standard part in VHDL, while allowing the user not only the choice of speed variations provided by the manufacturer, but the ability to enter his or her own timing values through **generic** variables. The examples provided in electronic form with the chapter illustrate the technique in detail. The proposed methodology will minimize library size and optimize simulation execution speed. An implementation using such a methodology can be seen in Figure 10.

This methodology requires the user to modify the **component instance specification** in the design every time a new timing value is required, as opposed to editing a data file, but most systems allow such changes to occur through the **configuration specification** mapping mechanisms, which introduce little overhead to the execution of a VHDL simulation.

The proposed methodology requires one generic parameter, called part_type, which provides the name of the type to be used for a given instance of the part. Two constant arrays are used to retrieve the timing values: the first one, called part_types, is an array of names of the available speed variations for the model. The second is an array of values, arranged by part type. Both support default values and allow the user to input his or her specific values. To facilitate functional modeling default timing values are set to zero delay.

```
entity STD86 is
    generic ( part_type: string: = "DEFAULT";
            tplh_low: tp_values : = (0 ns,0 ns,0 ns);  -- unit delay
            tphl_low: tp_values : = (0 ns,0 ns,0 ns);  -- unit delay
            tplh_high: tp_values : = (0 ns,0 ns,0 ns);  -- unit delay
            tphl_high: tp_values : = (0 ns,0 ns,0 ns));  -- unit delay

    port (A1,B1,A2,B2,A3,B3,A4,B4: in logic7;
        Y4,Y3,Y2,Y1: out logic7);
    constant part_types: pm18: =
        ("SN5486           ",
         "SN54LS86A          ",
         "SN7486           ",
         "SN74LS86A          ",
         "DEFAULT          ");
    constant spec_time: tv18: =
        ((5.25 ns,15 ns,23 ns),(3.85 ns,11 ns,17 ns),   --SN5486
         (6.3 ns,18 ns,30 ns),(4.55 ns,13 ns,22 ns),
         (4.2 ns,12 ns, 23 ns),(3.5 ns, 10 ns, 17 ns),  --SN54LS86A
         (7 ns, 20 ns, 30 ns),(4.6 ns, 13 ns, 22 ns),
         (5.25 ns,15 ns,23 ns),(3.85 ns,11 ns,17 ns),   --SN7486
         (6.3 ns,18 ns,30 ns),(4.55 ns,13 ns,22 ns),
         (4.2 ns,12 ns, 23 ns),(3.5 ns, 10 ns, 17 ns),  --SN74LS86A
         (7 ns, 20 ns, 30 ns),(4.6 ns, 13 ns, 22 ns),
         tphl_low,tphl_low,tplh_high,tphl_high);      --DEFAULT
    constant cpt:part_names: = pad(part_type);
    constant ctplh_low:tp_values: = spec_time(pick_tpd(cpt,part_types));
    constant ctphl_low:tp_values: = spec_time(pick_tpd(cpt,part_types)+1);
    constant ctplh_high:tp_values: = spec_time(pick_tpd(cpt,part_types)+2);
    constant ctphl_high:tp_values: = spec_time(pick_tpd(cpt,part_types)+3);
end STD86;
```

FIGURE 10 Standard Component Entity Declaration

Since **constants** are used, all the overhead takes place at elaboration time, and execution of the simulation has no speed degradation due to the methodology. Those systems which support incremental simulation will experience even less overhead at elaboration time since only those values that have changed will supposely be re-elaborated.

A model can be included in a design description in a simple and elegant manner, as shown by the **component instantiation** and **specification** in Figure 11. By using named association, only the parameters requiring values not yet specified in the model need to be specified by the user, thus gaining ease of use without sacrificing flexibility.

```
--
-- Preceeding VHDL code
--
U31: STD86
    generic map (part_type = > "SN54S86")
    port map ( -- usual port name associations
        );
--
-- Rest of design
--
```

FIGURE 11 Standard Component Instantiation

7. Toward a Standard

In working toward a standard set of modeling practices for the development of interoperable models of standard parts, the VHDL user community must resist the temptation to codify implementation styles. Doing so limits the creativity of implementors and stiffles further development. We should be concerned with identifying and explaining requirements, as well as evaluating methodologies for efficiency, and as a result, develop a set of guidelines which, although describing the external interfaces of the models, still allow complete freedom of implementation within the model or the **subprograms** in support **packages**.

The challenge is to understand the problem of model interoperability before we codify a standard, while at the same time not wait so long that existing commercial market forces become so strong as to make a standard futile. A standard would have to cover not just models, but support **packages** as well. Such a **package** must be complete, and must produce deterministically accurate results when used incorrectly as well as when properly used. Note that an error message is a deterministically accurate result if the same erroneous use will always generate the same error message. An indication of the difficulty of the problem faced by a standard making body can be obtained by reviewing some of the existing public domain **packages**.

When developing a standard **package** one must insure that no side effects are created when a function is implemented. The detection of a rising or falling edge in a signal waveform in **packages** is used to illustrate some points which need to be taken into consideration when developing **packages** for third party use.

Note that the interpretation problems described below arise not from implementation mistakes, but from a lack of communication due to insufficient documentation. Therefore the **package** used to implement the basic functions and procedures in a library must always be delivered in source form. The user should study the code provided, in order to understand the assumptions made by the implementors. There are as yet very few **packages** which have been developed and delivered to the public domain. Each was developed for a specific purpose and contributed many useful ideas, both in contents and style. Style and methodology implied in the implementations are as important as the contents.

The **package** provided by Vantage Analysis Systems contains two symmetrical **functions** (**f_rising_edge** and **f_falling_edge**) to allow the user to detect logic transitions. The description of the **functions** in the user manual, including examples of actual use, leads one to believe that the edge will be properly detected if the **function** is called with the correct syntax. The implementation of the **function** is a highly efficient table lookup which, in the case of the rising edge, returns the value TRUE if the signal has a state value of '1', and FALSE otherwise. The falling edge **function** is the symmetrical opposite.

The documentation provided with the **package** makes no mention of the fact that the functions provided do not check for an **event** on the signal in question when invoked, and thus may return an incorrect edge indication if called without such test being performed. A user may resonably assume that such a test is performed inside the **function**, while the implementors have assumed that the **function** would be invoked only after an event has been detected on the signal in question. As one can see, both assumptions are reasonable, yet taken together they lead to incorrect modeling of the part's behavior.

Both this **package** and the one provided by Zycad define a transition from an unknown state to either a high or low state as an edge. Since Verilog uses the same definition of edge transition, such an implementation is reasonable. Yet a user can logically expect that no edge will occur in a transition between an unknown and a logic high or low state since, by definition, an unknown is either high or low. Once again two reasonable assumptions can lead to incorrect modeling of electrical behavior when taken together.

Using a third party **package** to develop models is efficient and promotes code reusability. It is also froth with danger if the user cannot thoroughly understand not just its implementation but also the reasoning behind it. Therefore if the user is allowed to add models to an existing library, the source code for the underlying **package** of utility **subprograms** must be provided.

8. Summary

The opinions and examples described in this chapter should be helpful to those readers who are considering the implications of producing models which operate in more than one VHDL implementation, and with models developed by more than one vendor. Highlighted are issues related to modeling requirements over and above the needs met by a standard language.

Further study of the semantics of modeling electrical behavior using the VHDL syntax is required. A number of committees are presently active in this field. In general their membership is open to all, and additional contributors are always welcome. Readers interested in becoming involved, or in need of further details on planned or ongoing activities, can contact the IEEE standards organization, the VHDL Users' Group, or the EIA for information.

Chapter 7

Anomalies in VHDL and How to Address Them

Ken Scott

Synopsys, Inc.
Mountain View, California

This chapter discusses some of the anomalies inherent in VHDL which are commonly encountered when using the language. By 'anomalies' we mean obstructions which impede your ability to use the language to design or model useful circuits.

Presented first are common misconceptions about the language with clarifications on how the language really works in these situations.

Next some inconsistencies in the language are uncovered along with work arounds for them.

The ultimate aim of this chapter is to expose you to a number of potential problems in using the language, and to provide clarifications and guidance on how to work around them.

7.1   Common Misconceptions about VHDL

When designers are first introduced to using VHDL, they come from a variety of backgrounds which color their initial perceptions of the language. Some may have had some experience with modeling in another HDL, such as Zycad's ISP' or Silvar-Lisco's HHDL. Others may have had some experience with procedural programming languages. Still others may have had no experience with modeling whatsoever, having used graphical design techniques only.

It is often the case that some fundamental concept behind VHDL is misunderstood and, as a result, the designer is confounded when unable to get even relatively simple models working properly. Included in this chapter are some of the more commonly misunderstood concepts in the language. After reading this section, you will be able to establish a modeling methodology which will enable you to avoid making mistakes in common modeling situations.

7.1.1   VHDL Processes and Drivers

The most misunderstood concepts in VHDL are processes and drivers. To clarify them, the terms will first be defined. We'll then add detailed examples to illustrate the concepts behind them.

A *process* in VHDL is akin to a behavioral black-box or "gate" element in previous logic simulators. The process has inputs to which it is sensitive, and has output signals which it drives. The term 'process' here refers to any of the behavioral concurrent statements in VHDL. These include:

- Process statements.

- Concurrent assertion statements.

- Concurrent signal assignment statements.

- Concurrent procedure calls.

Processes are the fundamental building blocks in VHDL. The concurrent statements which have not been listed above are provided merely as a means of grouping together processes.

When a process includes a signal assignment, exactly one *driver* is created in that process for the signal. The driver is created to hold the contribution which the process is making to the value of the signal. The collection of drivers from one or more processes for any signal determines the resulting value on the actual signal.

This is illustrated with the example of a full-adder in figure 7.1.

```
package gates is
    procedure xor2( signal a,b : in bit; signal z : out bit );
    procedure or2( signal a,b : in bit; signal z : out bit );
    procedure and2( signal a,b : in bit; signal z : out bit );
end gates;

entity full_adder is
    port(
        a,b,cin : in bit;
        sum,cout : out bit );
end full_adder;

architecture subprogs of full_adder is
    use work.gates.all;
    signal s1, s2, s3 : bit;
begin
    p1: xor2( a,b,s1 );
    p2: xor2( s1,cin,sum );
    p3: and2( s1,cin,s2 );
    p4: and2( a,b,s3 );
    p5: or2 ( s2,s3,cout );
end subprogs;
--
```

Figure 7.1.   Procedural Full-Adder Model

The concurrent procedure calls labeled p1 through p5 can be thought of as components in a netlist, having inputs and generating outputs.

These concurrent procedure calls create drivers for the signals 's1', 'sum', 's2', 's3', and 'cout'. Since all of these signals are driven by only one process each, their values are determined by the respective concurrent procedure calls which drive them. They behave just as if they were driven by gates in a netlist.

Another way of describing the very same behavior involves using the VHDL component instantiation statement, as in figure 7.2.

```
package gates_comp is
    component xor2 port( a,b : in bit; z : out bit );
    end component;
    component and2 port( a,b : in bit; z : out bit );
    end component;
    component or2 port( a,b : in bit; z : out bit );
    end component;
end gates_comp;

architecture structural of full_adder is
    use work.gates_comp.all;
    signal s1, s2, s3 : bit;
begin
    c1: xor2 port map( a ,b ,s1 );
    c2: xor2 port map( s1,cin,sum );
    c3: and2 port map( s1,cin,s2 );
    c4: and2 port map( a ,b ,s3 );
    c5: or2 port map( s2,s3 ,cout );
end structural;
- -
```

Figure 7.2.   Structural Full-Adder Model

The purpose of showing these two different but quite similiar descriptions is to illustrate the point that in VHDL, processes behave as if they were behavioral components. They have individual contributions to the signals they drive, just as if they were gates themselves.

This differs significantly from other modeling environments where the component is considered to be indivisible. Processes within the component contribute equally to the overall component's behavior, but do not have separate contributions. Therefore there would never be more than one driver for a signal inside of a component in these modeling environments. Environments which obey this paradigm include Mentor's BLM, Zycad's ISP', Silvar-Lisco's HHDL, and Valid's UCP modeling systems.

Let's illustrate this with two examples of a register. The first example shown in figure 7.3 will not work properly. In the architecture 'doesnt_work' we have partitioned the function of the register into two processes. One process is responsible for clocking in new data (when the reset line is inactive), while the other is responsible for resetting the data back to zero. Why won't this work? Because there are two separate processes driving the same signal, and there is no resolution function associated with the 'data_out' signal to pick out the appropriate value. A VHDL compiler should report an error.

An architecture which will correctly implement the register is shown in figure 7.4. This second architecture works because there is only one process which assigns to the signal 'data_out', so only one driver gets created. Inside the process either the logical process 'do_clock' or the logical process 'do_reset' is performed, but never both at the same time.

To summarize, VHDL processes act like components in a netlist. Each has its own contribution to the signal(s) it drives; just like a component. Processes are NOT to be used to break up a behavioral model into logical sub-functions, unless each process is intended to drive a different set of signals or the signals are intended to be resolved. If you really need concurrent processes and they need to share state which behaves like a global variable, then see section 7.1.3.

7.1.2   Initialization of Signals

VHDL is a *hardware* description language. Its purpose is to describe the concurrent behavior that will be exhibited in electrical circuits. Signals in VHDL are inherently different from the storage types found in most programming languages because of this need to describe concur-

```
entity register_4 is
    port(
        clock,reset : in bit;
        data_in : in bit_vector( 0 to 3 );
        data_out : out bit_vector( 0 to 3 ) );
end register_4;

architecture doesnt_work of register_4 is
begin
    do_clock:
    process( clock )
    begin
        if ((clock = '1') and (reset /= '1')) then
            data_out <= data_in;
        end if;
    end process do_clock;

    do_reset:
    process( reset )
    begin
        if (reset = '1') then
            data_out <= B"0000";
        end if;
    end process do_reset;
end doesnt_work;
- -
```

Figure 7.3.   Incorrect Model for a 4-bit Register

rency.

The fundamental storage type in most programming languages is the *variable*. Variables are supported in VHDL as well. When a VHDL variable is assigned, it takes on the value immediately. New assignments to a variable merely supply new values for it which overwrite the previous value. In VHDL, variables are usually initialized in the place where they are declared. For example,

variable i : integer := f(now) * 16;

defines a variable named 'i' and assigns it the initial value returned by the function call 'f' with parameter 'now' multiplied by 16.

```
architecture WORKS of register_4 is
begin
   handle_both:
   process( clock, reset )
   begin
      - - do_reset
      if ((reset = '1') and (reset'event)) then
         data_out <= B"0000";

      - - do_clock
      elsif ((reset = '0') and (clock = '1') and (clock'event)) then
         data_out <= data_in;
      end if;
   end process handle_both;
end WORKS;
- -
```

Figure 7.4.   Correct Model for a 4-bit Register

VHDL signals, on the other hand, are driven by potentially many different processes at the same time. The value of a signal at any point in time is calculated by arbitrating between all of the driving values for that signal and choosing an appropriate value. Because of this fundamental difference both the assignment and the initialization of signals are different than the assignment and initialization of variables. In order to understand exactly how signals are initialized it will help to introduce a new concept, the VHDL source, and explain how sources affect the initialization process and how they, in turn, can be initialized.

VHDL Sources and Initialization

A *source* in VHDL contributes to a signal's value. The sources of a signal can either be process drivers (as discussed earlier) or component ports of mode 'out', 'inout', or 'buffer' with which a signal has been associated. The collection of all sources is used when determining the value of a signal at any point in time. A contribution by a component port is termed a non-driver source.

The initial value of a signal is computed using the collection of its sources. The initial values of the sources for a signal are computed and are then used to initialize the signal. The key to correctly initializing a VHDL signal is to correctly initialize all of its sources.

VHDL provides an explicit mechanism for specifying the *default value* for a signal and for its process driver sources. This mechanism does not apply to the component non-driver sources. Only when a signal has no sources is the default value used as the initial value of the signal. If a signal has process driver sources then this default value is used as the initial value for these sources. For example, the following declaration will cause all process driver sources of the signal 'flag' to be set to '0':

signal flag : bit := '0';

The syntax used to specify the default value for a signal is identical to the syntax used to specify the initial value for a variable. The key difference between the two is that *the default value specified for the signal does not necessarily define the initial value of the signal.* The two examples shown in the following section will illustrate situations in which the default value specified for a signal is not the same as the value to which the signal gets initialized.

Multiple-Driver Signal Initialization Example

In this first example, shown in figure 7.5, a circuit is described which contains two processes which are simultaneously driving the output signal [1] 'flag'. The signal 'flag' has been declared to be of type 'mvl', which we'll assume is a 4-valued logic type ('0','1','x','z'). It also has associated with it a resolution function WIRE_X which is a function which emulates a bus. (At most one source of a signal can have a value other than the high-impedance value 'z'; otherwise there's a bus error and the signal will be assigned the unknown value 'x'.) The signal 'flag' finally

---

[1]In VHDL, a port is a form of signal.

has specified a default value of '0' which will be applied as the initial value for all of its driver sources.

```
entity test is
    port( flag : out WIRE_X mvl := '0' );
end test;

architecture twodrivers of test is
begin
    p1: flag <= some_func;
    p2: flag <= some_other_func;
end twodrivers;
- -
```

Figure 7.5.   Multiple Drivers Driving Signal

Since the default value for the signal 'flag' is '0', this will be the initial value assigned to the drivers associated with processes p1 and p2. During initialization the resolution function will be called with both drivers attempting to drive the bus to a '0' value. As a result, there will be a bus contention error and the value 'x' will be assigned to the signal 'flag'. This makes sense from an electrical point of view, as the initialization of the wire in the real circuit must follow a more elaborate protocol than just turning on all its drivers at once. This demonstrates one situation in which the default value specified for the signal does not ultimately indicate the initial value for that signal.

Example with Non-Driver Source Signal Initialization

In the next example, the VHDL from figure 7.5 has been modified so that instead of two processes driving the signal 'flag' there is now one port of mode 'out' which is the sole source for the signal. This modified example is shown in figure 7.6.

In this case, the default value associated with the signal 'flag' will have *no effect whatsoever* on the initial value of the signal, because the

```
entity test is
    port( flag : out WIRE_X mvl := '0' );
end test;

architecture oneport of test is
    component drive_comp
        port ( o : out mvl );
    end component;
begin
    c1: drive_comp port map( flag );
end oneport;
- -
```

Figure 7.6.   Single 'Out' Mode Port Driving Signal

default value specifies only the initial value for process driver sources. The component instance 'c1' supplies a non-driver source for the signal 'flag', so the initial value of that source is defined (not surprisingly) by the entity which ultimately is associated with the component instance.[2] Again this makes sense from an electrical standpoint, as any component which matched the interface of 'drive_comp' could be associated with the instance 'c1'. These components could have different initial values with which they drive.

Once again this demonstrates that the signal initialization mechanisms defined by VHDL make sense from an electrical point of view, and at the same time do not always match up with the assumption that the default value associated with a signal will define its initial value.

7.1.3   Working Around the Lack of Global Variables in VHDL

Global variables are useful for sharing information between processes without having to describe the exact electrical details of how such infor-

---

[2]Note, you could also give a default value in the component declaration. This value is overriden by the configured entity source value, if it exists.

mation is to be shared in a real circuit. They are not allowed in VHDL because there are ways they can be used which will cause unpredictable simulation results. For example, if two processes were scheduled to execute at the same time and one wanted to assign to a global variable while the other wanted to read it, there would have to be a rule specifying which process gets executed first. Otherwise this model might simulate differently on different VHDL simulators.

Nevertheless global variables would be useful in certain modeling situations. An example is shown in figure 7.7 indicating how global variables might be used to help implement a counter with an asynchronous reset line. Unfortunately, the example is illegal VHDL.

In this example either the first process will be setting the value of 'count' back to zero, or the second process will be incrementing it. Because of the way they've been modeled, it will never be the case that both processes will try to assign to the variable at the same time. In this simple modeling situation global variables would be helpful. The obvious way to avoid the need for global variables is to put all the behavior in one process. This was shown in figure 7.4 earlier. But when this is not possible because true asynchronous behavior is needed, there are other methods of implementing a similar behavior.

Use Intermediate Signals

One way to work around the lack of global variables is to use intermediate signals and a conditional signal assignment statement. The idea is to make use of the fact that both processes will not want to assign to 'count' at the same time. Therefore, construct a signal and associated process to emulate this behavior. A modified example containing a legal architecture for the entity counter is shown in figure 7.8.

In this example, the first process assigns to the intermediate signal 'resetcount', while the second process assigns to the intermediate signal 'incrcount'. The signal 'count' will be assigned the value driven by the first process whenever the intermediate signal associated with this pro-

```
- - Note that this example is illegal vhdl!
entity counter is
    port(
        clk,reset : in bit;
        data : out integer );
end counter;

architecture behavioral1 of counter is
    variable count : integer := 0; - - illegal!
begin
    process( reset )
    begin
        if (reset = '1') then
            count := 0;
            data <= count;
        end if;
    end process;
    process( clk )
    begin
        if (reset = '0') and (clk = '1') then
            count := count + 1;
            data <= count;
        end if;
    end process;
end behavioral1;
- -
```

Figure 7.7.   Illegal Use of Global Variables

cess is assigned. 'Count' will take on the value of 'incrcount' when its value has been assigned.

This modeling solution provides a clean and general way to allow two processes to cooperate in driving an output signal. The conditional signal assignment statement also provides a good way to prioritize the contributions of various processes if they happen to execute at the same time. The drawback is the need to declare and keep track of extra signals, which results in a less efficient simulation.

```
architecture behavioral2 of counter is
    signal resetcount, incrcount, count : integer := 0;
begin
    process( reset )
    begin
        if (reset = '1') then
            resetcount <= 0;
        end if;
    end process;
    process( clk )
    begin
        if (clk = '1') and (reset = '0') then
            incrcount <= count + 1;
        end if;
    end process;

    count <= resetcount when (not resetcount'quiet) else
             incrcount when (not incrcount'quiet) else
             count;

    data <= count;
end behavioral2;
- -
```

Figure 7.8. Using Intermediate Signals instead of Variables

Use VHDL Registers

An approach which is quite similiar to the one used in the previous example is to declare the intermediate signal 'count' to be a signal of kind *register*. Register signals are a special type of signal which retain their last value when all process drivers have been assigned the predefined value **null**.

We will need a resolution function for the following example. It is shown in figure 7.9. This resolution function will be associated with the integer register signal 'count' and will report an error if at any time after 0 ns the number of drivers actively contributing to the value of 'count' is greater than one. If one and only one driver is active, the resolution function will return the value of that driver. If neither driver is active then the resolution function will not be called and the signal will retain its last value (because it is a register).

```
function resolved( s : in int_array ) return integer is
begin
    if (s'length > 1) then
        if (now = 0 ns) then
            return s(s'left);
        end if;
        assert false report "Bus contention on signal count.";
        return 0;
    end if;
    return s(s'left);
end;
- -
```

Figure 7.9.   Resolution Function for Resolving Registers

The new example architecture is shown in figure 7.10. At time zero, the signal 'count' will be set to 0 because the resolution function treats time 0 ns as a special case. Whenever the 'reset' signal is active the 0 value will be assigned to the signal 'count'. Otherwise, if the 'clk' signal is active and 'reset' is inactive then the value of the signal 'count' will be incremented.

This modeling solution is useful in situations when multiple processes wish to share data but an error condition should occur if any of the processes attempt to assign to the common data at the same time. In this particular example we've precluded that possibility by disabling the increment operation when the reset line is inactive.

All of these solutions are reasonable ways to share data amongst multiple logical processes. Which solution you use depends upon the modeling situation.

```
architecture behavioral3 of counter is
   signal count : resolved integer register := 0;
begin
   process( reset )
   begin
      if (reset = '1') then
         count <= 0, null after 1 fs;
      else
         count <= null; - - for initialization
      end if;
   end process;
   process( clk )
   begin
      if (reset = '0') and (clk = '1') then
         count <= count + 1, null after 1 fs;
      else
         count <= null; - - for initialization
      end if;
   end process;

   data <= count;
end behavioral3;
- -
```

Figure 7.10.   Using VHDL Registers instead of Variables

7.1.4   Use of 'out' and 'buffer' Mode Ports

Ports of mode 'out' and 'buffer' have similiar characteristics. Yet there are modeling situations in which one is more appropriate than the other. This section will define the differences between the two modes and offer suggestions about when each should be used.

Ports of mode 'out' drive the actual signal to which they are connected. There are no restrictions on the type of network which can be driven by an 'out' port, nor are there any restrictions on how an 'out' mode port may be driven. Ports of mode 'out' cannot have their values read internal to the entity in which they've been declared for reasons explained later.

Ports of mode 'buffer' also drive the actual signal to which they are connected. There is a restriction that a buffer mode port may be driven by at most one source and that they drive networks which contain only one source. Ports of mode 'buffer' may have their values read in the entity in which they've been declared, unlike ports of mode 'out'.

In order to better understand the restrictions and capabilities of these two port modes, it will help to gain an understanding of how signal values in VHDL are propagated.

Signal Propagation in VHDL

When a port drives a network, it is considered to be a source for that network. In other words, it contributes a value to the network which has an influence on the value at which the network will stabilize. The value which a port contributes to the network it drives is called its *driving value*. The driving value of a port is not accessible in VHDL. It cannot be read by a model.

When a VHDL simulator evaluates a signal, all of the sources for that signal are examined and the actual value of the signal is computed. This is the *effective value* for the signal, which is seen when a signal is read.

Ports of mode 'out' are not allowed to have their values read because of two reasons. The first is that if the effective value of an 'out' mode port were defined to be the same as the effective value of the network it drives then there would be no difference between an 'out' mode port and an 'inout' port. Second, the driving value of a port is not a meaningful value unless there is only one source contributing to that driving value. When more than one process is contributing to the value of an 'out' port, the 'driving value' of the port is the result of combining the values with which the processes are driving and resolving them. In each process what is often desired is to know the value with which the 'out' port is being driven *from within that process*.

The 'buffer' mode port is an 'out' port which has been sufficiently restricted so that the effective value of the port is also, by definition, the same as the driving value of the port. Since there can be at most one source for the port, it is assured that the driving value of the port is the result of the sole contribution of the sole source. Since the port can be the only source for the actual signal to which it is attached, it is insured again that the effective value on that outside signal will be the same (under normal circumstances) as the driving value of the port. [3]

When to use 'buffer' / 'out' mode ports

In general, the following guidelines are useful when deciding whether an output port should be declared mode 'out' or mode 'buffer':

- When modeling ports which can never be wired together use *buffer* mode ports.

- When modeling ports such as tristate or open collector / emitter ports which can be wired together use *out* mode ports. Since ports of this mode cannot be read, the next section will describe work arounds for reading 'out' mode ports.

Work arounds to allow reading 'out' mode ports

In those situations in which a model must be written which cannot conform to the restrictions implied by 'buffer' mode ports, there are ways to allow the driving value of an 'out' port to be read.

The first involves the introduction of a temporary signal. The idea is to insert an internal signal between the 'out' port and its drivers. An example of this is shown in figure 7.11. The value with which 'out_sig' is being driven by this process can then be read directly from the intermediate signals. There is an additional simulation cycle of delay introduced before the output signal will propagate to the proper value, but in most modeling situations this is not a problem.

---

[3]The only exception to this rule is when conversion functions are attached to the port or the port is of a resolved type and the resolution function performs some non-identity transformation on the driving value.

```
entity test is
    port( clock : in bit; out_sig : out bit );
end test;

architecture read_out of test is
    signal out_tmp : bit;
begin
    process( clock )
    begin
        out_tmp <= not out_tmp;
        - - Can read driving value we contribute to out_sig.
    end process;
    out_sig <= out_tmp;
end read_out;
- -
```

Figure 7.11.   Using a Temporary Signal to Read Driving Value

A second solution involves the use of variables when the signal driver is internal to a process, as in the previous example. This is shown in figure 7.12. The variable keeps track of the most recent value used to drive the output signal. This solution works well when modeling with zero delay signal assignments. When nonzero or inertial delay assignment is utilized, however, it is likely that the variable will not track the value of the signal at all times.

7.1.5   Use of Bus and Register Signals

Bus and register signals were introduced into VHDL to provide a mechanism in the language so that signal types would not have to include a high-impedance state in order to be bussed. Predefined types BIT, INTEGER, and REAL do not have high-impedance states but make sense in a bussing scheme where only one source is meant to drive the bus at a time. A package which would resolve type BIT according to this bussing scheme is shown in figure 7.13.

An example which declares a signal of this resolved type to be of kind

```
architecture use_variables of test is
begin
    process( clock )
        variable out_int : bit;
    begin
        out_sig <= not out_int;
        out_int := not out_int;
    end process;
end use_variables;
- -
```

Figure 7.12.   Use of Variable to Store Driving Value

```
package resolve_bit is
    function bus_resolve( s : bit_vector ) return bit;
    subtype rbit is bus_resolve bit;
    type rbit_vector is array( natural range <> ) of rbit;
    subtype rbit4 is rbit_vector( 3 downto 0 );
end resolve_bit;

package body resolve_bit is
    function bus_resolve( s : bit_vector ) return bit is
    begin
        if (s'length = 0) then
            assert (false) report "zero sources driving bus";
            return '0';
        elsif (s'length /= 1) and (now /= 0 ns) then
            assert (false) report "multiple sources driving bus";
            return '0';
        else
            return s(s'left);
        end if;
    end bus_resolve;
end resolve_bit;
- -
```

Figure 7.13.   Package Defining Resolution Function for type BIT

'bus' and then shows how the VHDL code could be written to control
the bus is shown in figure 7.14.

```
use  work.resolve_bit.all;
entity micro_p is
    port(
        control_bus : in bit;
        bus_sig : inout rbit4 bus );
end micro_p;

architecture example of micro_p is
begin
    - - Drives bus to all zero's when control signal is '0'.
    driver0:
    block(control_bus = '0')
    begin
       bus_sig <= guarded b"0000";
    end block driver0;

    - - Drives bus to all one's when control signal is '1'.
    driver1:
    block(control_bus = '1')
    begin
       bus_sig <= guarded b"1111";
    end block driver1;
end example;
    - -
```

Figure 7.14.   Using Registers to Model the Behavior of a Bus

The guarded signal assignment statements are guarded by mutually exclusive conditions, such that one of them will always be turned on while the other will always be turned off. Thus, we've described the behavior of a bus without having to use a type with a high- impedance state.

The problem in using signal kinds **bus** and **register** arises from the fact that they no longer work if you attempt to introduce hierarchy into the design. For example, if ever the blocks 'driver0' or 'driver1' got so complex that it made sense to encapsulate them in their own entity/architecture, it would be impossible to continue to use a type without a high-impedance state. This is because VHDL process drivers may be turned off by guarded signal assignments, but the sources resulting from connecting the signal 'bus_sig' to an 'out', 'inout', or 'buffer' mode port can NEVER be turned-off. Thus, registers and busses can only be used within one level of hierarchy.

With that strong of a restriction placed on their use, the designer does not really gain much by learning how to use them because once the process of decomposing the design has begun they must be converted back to signals declared with a type which has a high-impedance state. The suggested methodology then is to use tri-state logic exclusively and not use bus and register signal kinds. Note that the signal kinds are sometimes useful for implementing global variables, as shown in a previous section, but they are not very useful for describing busses. Also note that the guarded assignment used in the block statements is independent of the concept and use of guarded signals.

7.1.6   Predefined Signal Attributes

VHDL supplies a number of predefined attributes which can be used with signals to obtain information about their history. Both the 'event and the 'stable attributes can be used to provide information about when a signal was last updated.

The 'event attribute is a *function* attribute whose return value is of type boolean. It will return the boolean value TRUE whenever there has been an event on the signal with which it is associated at the beginning of the current simulation cycle. It returns FALSE if there has not been an event during the current cycle. When associated with a signal it can be used anywhere a function call can be used.

The 'stable attribute is a *signal* attribute whose value is also of type boolean. The value of this signal will be TRUE whenever (in the absence of a time expression) there has not been an event on the signal with which it is associated during the current simulation cycle. It will be FALSE if there has in fact been an event during the current cycle. When used without a time expression, it is functionally the inverse of the 'event attribute. It implies a signal, however, and as such can be used anywhere a signal can be used in VHDL.

Although these two attributes are closely related, they are not always interchangeable. There are situations in which it makes sense to use only one of these attributes, and the following sections will discuss a few of those situations.[4]

Signal Attributes in Guarded Blocks

A *guarded block* in VHDL is a construct commonly used to model control line-sensitive behavior. In an edge-sensitive D flip-flop, for example, we want to transfer the data from the input line (D) to the output lines (Q and QB) only when a positive (or negative) edge is detected on the clock. In all other cases, we want the outputs to remain constant.

```
bgood:block ((clk = '1') and (not clk'stable))
begin
    Q <= guarded D;
    QB <= guarded not D;
end block bgood;
```

Note that the *'stable* attribute has been used in the guard expression to indicate that the guarded signal assignment statements within the block should only be activated when the 'clk' signal has just made a transition to '1'.

Since the 'stable and the 'event attributes are functionally opposite, a common mistake that is made is to believe that substituting the clk'event attribute into this expression for the expression (not clk'stable) will cause the same behavior.

---

[4]Note that the *'active* and the *'quiet* attributes have the same characteristics as the 'event and the 'stable attributes, respectively. Therefore, the following sections apply to these attributes as well.

```
bbad:block ((clk = '1') and (clk'event))
begin
    Q <= guarded D;
    QB <= guarded not D;
end block bbad;
```

This is not the case, however, because of the way in which guard expressions get updated. *Every time a signal in a guard expression is activated the guard expression gets re-evaluated.*

In the block 'bgood' when the 'clk' signal rises the guard expression will be evaluated (because of the activity on the signal 'clk') and the resulting value will be TRUE. In the cycle immediately after, the clk'stable signal will change value from FALSE to TRUE which will cause the guard expression to again be evaluated and be reset to FALSE. This is the desired behavior.

In the block 'bbad' the 'clk' signal rising causes the guard expression to be evaluated and the resulting expression to be TRUE. In the cycle immediately following this transition, however, there are no signals in the guard expression which are active! The 'event attribute, you'll recall, implies a function call and not a signal. Because of this, the guard expression in block 'bbad' will not be reset until the signal 'clk' has some activity on it. The end result is that the D flip-flop of 'bbad' becomes a latch, propagating the input 'D' to the outputs 'Q' and 'QB' as long as the 'clk' signal remains high.

This is clearly a situation in which the 'event attribute and the 'stable attribute are NOT interchangeable. The 'stable attribute should be used whenever the modeling situation calls for some behavior which is exhibited only by signals.

Simulation Efficiency Considerations for Signal Attributes

It is often the case when constructing processes which are sensitive to more than one signal that there needs to be a way to distinguish *which* signal in the sensitivity list caused a process to be invoked.

A simple example of this is shown in figure 7.15, which contains the VHDL for a 4-bit register with an asynchronous reset line. In this example we need to be able to detect a rising-edge on either the 'reset' line or on the 'clock' line.

```
entity register4 is
    port(
        clock,reset : in bit;
        data_in : in bit_vector( 0 to 3 );
        data_out : out bit_vector( 0 to 3 ) );
end register4;

architecture behavioral of register4 is
begin
    process( clock, reset )
    begin
        if ((reset = '1') and (reset'event)) then
            data_out <= B"0000";
        elsif ((reset = '0') and (clock = '1') and (not clock'stable)) then
            data_out <= data_in;
        end if;
    end process;
end behavioral;
- -
```

Figure 7.15. 4-bit Register with Asynchronous Reset

In the case of the 'reset' line, we've used the predefined attribute 'event. In the case of the 'clock' line we've used the predefined attribute 'stable. In both cases the same functionality is being achieved. Which construct can be simulated more efficiently?

The 'stable attribute is the more general of the two, in that it can be used as a signal in a sensitivity list, a guard expression, or in a port map clause. It can also provide more information than the 'event attribute when supplied a time argument T; 'stable can be used to tell if the signal changed value in the last simulation cycle or if the signal changed value in the last T units of time.

Because of the more general functionality of 'stable, a VHDL simulator in general will have to do more processing to keep track of the attribute. It is possible for an implementation to recognize the special case where 'stable is being used in an expression as if it were 'event, but in general 'stable cannot be simulated more efficiently than 'event. For that reason, the best strategy from a simulation efficiency point of view is to use the 'event attribute instead of the 'stable attribute whenever possible.

The previous sections outlined issues and problems with using the language as defined. The following looks into problems with the language itself and the proposed resolutions.

7.2   VHDL Language Inconsistencies

Since the VHDL 1076-1987 standard was ratified, work has been done by the VHDL Analysis and Standardization Group (VASG) on identifying and resolving inconsistencies in the language definition. The intent of this section is to enumerate those language issues which are likely to affect most designers using VHDL, and to suggest a methodology in each instance for working around the problem.

The reason why these inconsistencies are important is that they involve features of the VHDL language which are more likely to be nonportable across different toolsets than other features. Since most of the toolset vendors are active participants in the VASG, it's likely that adhering to the recommendations of this group when writing VHDL models will result in code which is more likely to work across VHDL systems.

7.2.1   The Textio Package

The most significant inconsistency in the VHDL language definition is in the predefined Textio package. The Textio package was intended to support formatted ascii I/O operations. The problem with the package was that the file format expected when using Textio to read a file or when using Textio to write out a file wasn't defined anywhere in the Language Reference Manual. Tool developers had no guidance on what these formats should look like, so different interpretations were implemented.

Type	Format	Example
BIT	Single character, either 1 or 0	1
BIT_VECTOR	Sequence of characters, either 1 or 0	1101
BOOLEAN	An identifier, either TRUE or FALSE	TRUE
CHARACTER	A single ascii character	A
INTEGER	Integer literal, as in LRM 13.4.1	34
REAL	Floating-point literal, also as in 13.4.1	3.14
STRING	Sequence of characters in the string	hello,world
TIME	Optional numeric literal, TIME unit	14.5 ns

Figure 7.16.   Ascii format used by package Textio

The VASG recognized this as a major issue and resolved to correct it by explicitly defining the file format which Textio would use. In this section we'll outline the file format proposed and accepted by the VASG. This format should be used when writing Textio input files, or expected when looking at files produced by the Textio write routines. The table shown in figure 7.16 summarizes the general ascii format expected for each of the possible predefined types that may be read or written using the Textio package.

The model shown in figure 7.17 can be used to test the full_adder design shown at the beginning of the chapter in figure 7.1. The ascii file which can be read by this model to test all possible input combinations for the full_adder design is shown in figure 7.18.

There are other inconsistencies in the VHDL definition which are worth mentioning. They are listed in the following sections.

```
entity tester is
    port (
        a,b,cin : out bit;
        sum,cout : in bit );
end tester;

use std.textio.all;
architecture use_textio of tester is
begin
    read_apply_check:
    process
        file stimulus : text is in "f7.18.stm";
        variable vector : line;
        variable value : bit;
        variable space : character;
        variable t : time;
    begin
        while (not endfile(stimulus)) loop
            readline( stimulus,vector );
            - - Format:
            - - ApplyTime, a,b,cin, WaitTime, expect_sum,expect_cout
            read( vector,t ); wait for (t - now);
            read( vector,space );
            read( vector,value ); a <= value;
            read( vector,value ); b <= value;
            read( vector,value ); cin <= value;
            read( vector,space );
            read( vector,t ); wait for t;
            read( vector,space );
            read( vector,value ); assert (sum = value);
            read( vector,value ); assert (cout = value);
        end loop;
        wait;
    end process;
end use_textio;
- -
```

Figure 7.17.   Test Bench for Full Adder from Figure 7.1

7.2.2   Spaces in Abstract / Physical Literals

The VHDL Language Reference Manual (LRM) requires that there be at least one space in between the numeric literal and the unit name in a physical literal. At the same time there are many examples in the LRM

```
10 ns 000 2 ns 00
20 ns 001 2 ns 10
30 ns 010 2 ns 10
40 ns 011 2 ns 01
50 ns 100 2 ns 10
60 ns 101 2 ns 01
70 ns 110 2 ns 01
80 ns 111 2 ns 11
<EOF>
```

Figure 7.18.   Stimulus File for Full-Adder

which do not conform to this rule.[5] For example at the end of section 10.3 in the LRM there is an example which contains the following signal assignment statement:

B <= A **after** 15ns;

This is illegal VHDL, as there must be a space between the numeric literal '15' and the time unit 'ns', as shown in the corrected statement below:

B <= A **after** 15 ns;

Including a space between the numeric literal and the unit name in a physical literal will insure that the models you write will be correct and portable.

---

[5]Examples and notes of the LRM are not part of the official standard. They are provided for clarification only.

7.2.3   Null Slices

A *null slice* is a slice of an array which contains no elements. In the VHDL LRM there are two ways in which a null slice can be created. The following example illustrates the two ways to create a null slice:

signal S : BIT_VECTOR(31 **downto** 0);

S(14 **downto** 15)
S(1 **to** 2)

The first slice is null because there are no elements included in the range (14 downto 15). The second slice is null because the direction of the slice ("to") is different than the direction of the signal being sliced ("downto").

When a slice with a null range is created, it must be possible to define the leftmost and rightmost indices and the direction of the slice in such a way as to indicate that there are no elements in the slice. There is a case when attempting to slice a one-element array where defining the indices and direction of a null slice is impossible. Because of this it has been recommended that it be made illegal for the direction of the slice to differ from the direction of the object (constant, variable, or signal) being sliced.

When writing VHDL models the direction of the slice should always match the direction of the object being sliced, in order for the code to be portable.

7.2.4   Slices in Case Statements

The LRM requires that the subtype of the select expression in a case statement be well-defined. An example where the subtype of the selection expression has been clearly defined is shown in figure 7.19. In this example the subtype of the case statement select expression is the type of the signal 'choice', which is 'bit'.

```
b:block
   signal choice : bit;
begin
   process( choice )
   begin
      case choice is
         when '0' => - - do something;
         when '1' => - - do something else;
      end case;
   end process;
end block b;
- -
```

Figure 7.19.   Simple Case Statement Example

There is a problem in the LRM in that the subtype of a slice expression is not defined. As a result, slices cannot be used as the select
expression in a case statement. The workaround for this oversight is to
explicitly qualify the slice expression with the subtype that represents
the sliced value. The example shown in figure 7.20 illustrates how this
can be done. In this example we are extracting the 8 high-order bits of
the 'instr' signal and using that to select an instruction to execute.

The ability to use slices in case statements is a very useful modeling
feature. To insure that such VHDL will be portable you should qualify
the slice expression with the proper subtype.

7.2.5   Resolution Function Parameters

Formal function parameters are allowed, in general, to be constants
or signals. If a signal is passed as a parameter to a function, then certain
predefined attributes of the actual signal passed may be read. These
attributes tell what the last value of the signal was ('LAST_VALUE),
whether the signal had any transactions ('ACTIVE) or events ('EVENT)
during the current simulation cycle, and how much time has elapsed
since the last transaction ('LAST_ACTIVE) or event ('LAST_EVENT)

```
process(instr)
   subtype opcode is bit_vector( 7 downto 0 ):
begin
   case opcode'(instr( 31 downto 24 )) is
       when b"11111111" => - - add;
       when b"11111110" => - - subtract;
           o
           o
           o
       when others => - - report an error;
   end case;
end process;
   - -
```

Figure 7.20.   Case Statement Using Slices

occurred on the signal.

A resolution function is a function. It is passed the values of the sources for a resolved signal and arbitrates between these values to determine what the resulting value of the signal should be. The sources which make up the single array input parameter may be component port sources or they may be VHDL process driver sources.

The LRM does not restrict the formal parameter of a resolution function from being a signal parameter. If the parameter to such a function were allowed to be a signal, however, the language would require that all of the predefined signal attributes listed previously be supported for that signal parameter.

The problem with this is that these attributes are not defined for process driver sources. There is no mechanism in the language for determining the last time a driver had an event on it, or for getting any of the other predefined signal attribute pieces of information. These mechanisms would have to be defined in order to allow resolution function parameters to be signals.

For this reason, the VASG recommended that signals not be allowed as formal parameters of a resolution function. The resolution function parameter should always be a constant.

7.3   Summary

Misunderstandings about the way VHDL works can result in wasted design time. This chapter has presented an overview of some of the more common misconceptions about the VHDL language, and also listed some of the areas in which its definition is inconsistent. This information should enable you to skirt around the more troublesome areas in the language, and as a result become more effective when using VHDL to do your design work.

7.4   References

[1] "IEEE Standard 1076-1987 VHDL Language Reference Manual," IEEE Inc., New York, 1988.

[2] Lipsett, R. et al., VHDL: Hardware Description and Design, Kluwer Academic Publishers, New York, 1989.

[3] Scott, K., "Modeling Guidelines for Efficient Simulation," VHDL Users' Group Fall 1989 Meeting, pp. 5-08 - 5-13, October, 1989.

[4] "The Sense of the VASG", VHDL Users' Group Fall 1989 Meeting, October, 1989.

Index